MEMS and Nanotechnology for Gas Sensors

MEMS and Nanotechnology for Gas Sensors

Sunipa Roy

Chandan Kumar Sarkar

CRC Press
Taylor & Francis Group
Boca Raton London New York

CRC Press is an imprint of the
Taylor & Francis Group, an **informa** business

CRC Press
Taylor & Francis Group
6000 Broken Sound Parkway NW, Suite 300
Boca Raton, FL 33487-2742

Printed on acid-free paper
Version Date: 20150803

International Standard Book Number-13: 978-1-4987-0012-2 (Hardback)

Visit the Taylor & Francis Web site at
http://www.taylorandfrancis.com

and the CRC Press Web site at
http://www.crcpress.com

MIX
Paper from
responsible sources
FSC
www.fsc.org
FSC® C013056

Printed and bound in Great Britain by
TJ International Ltd, Padstow, Cornwall

Dedicated to our families

for their

love and support

Contents

Preface xiii
Authors xvii

Section I Fabrication Procedure

1. Introduction 3
- 1.1 Cleanroom Technology 3
- 1.2 Microelectromechanical System 6
  - 1.2.1 History 7
  - 1.2.2 Definitions and Classifications 8
  - 1.2.3 Market and Application 9
  - 1.2.4 Materials for MEMS 10
- 1.3 Significance of MEMS 14
- References 15

2. Substrate for MEMS 17
- 2.1 Introduction 17
- 2.2 Silicon: The Base 18
  - 2.2.1 Silicon as a Semiconductor 18
  - 2.2.2 Surface Contamination 19
    - 2.2.2.1 Contaminants on Silicon Wafers 20
  - 2.2.3 Cleaning and Etching 22
    - 2.2.3.1 Cleaning 22
    - 2.2.3.2 Etching 22
- 2.3 Dielectrics 26
  - 2.3.1 Silicon Dioxide (SiO_2) 27
  - 2.3.2 Silicon Nitride (Si_3N_4) 28
  - 2.3.3 Low-Temperature Oxidation 29
  - 2.3.4 Oxide Properties 29
- References 30

3. Deposition 31
- 3.1 Physical Vapour Deposition 31
  - 3.1.1 Vacuum Technology for MEMS 32
  - 3.1.2 e-Beam Evaporation 34
  - 3.1.3 Thermal Evaporation 34
  - 3.1.4 Sputtering 36
  - 3.1.5 Molecular Beam Epitaxy 37

3.2 Chemical Vapour Deposition 38
3.2.1 Atmospheric Pressure CVD (APCVD) 38
3.2.2 Plasma CVD 39
3.2.3 MOCVD 40
3.3 Metallization 42
3.3.1 Different Types of Metallization 42
3.3.2 Methods 44
3.3.2.1 Filament Evaporation 45
3.3.2.2 Electron Beam Evaporation 45
3.3.2.3 Induction Evaporation 45
3.3.2.4 Sputtering 45
3.3.3 Wire Bonding 45
References 47

4. Photolithography: Pattern Transfer 49
4.1 Introduction 49
4.2 Photoresist for Structuring 50
4.3 Some Important Properties of Photoresist 50
4.3.1 Sensitivity 50
4.3.2 Adhesion 51
4.3.3 Etch Resistance 51
4.3.4 Bubble Formation 51
4.4 Types of Photoresists: Negative and Positive Photoresists 52
4.4.1 Negative Photoresists 52
4.4.2 Positive Photoresists 53
4.5 Designing of Mask Layout 53
4.6 Photolithography Process 54
4.7 Application of Photoresist and Prebake 54
4.8 Alignment, Exposure, and Pattern Formation 55
4.9 PR Developer and Postbake 56
4.10 Stripping (Photoresist Removal) 56
4.11 Some Advanced Lithographic Techniques 57
4.11.1 Electron Beam Lithography 57
4.11.2 Ion Beam Lithography 58
4.11.3 X-Ray Lithography 59
4.11.4 Phase-Shift Lithography 60

5. Structuring MEMS: Micromachining 63
5.1 Introduction 63
5.2 Bulk Micromachining 64
5.2.1 Wet Etching 65
5.2.1.1 Isotropic and Anisotropic: Empirical Observations 66
5.2.1.2 Convex and Concave Corner Compensations 69
5.2.2 Dry Etching 72

5.3 Surface Micromachining....72
5.3.1 Processes....72
5.3.2 Hurdles....74
5.3.3 Lift-Off versus Etch Back....74
5.3.3.1 Lift-Off....74
5.3.3.2 Etch Back....75
5.4 Etch-Stop Technique....76
5.4.1 Boron Etch Stop....77
5.4.2 Electrochemical Etch Stop....78
5.4.3 Photo-Assisted Electrochemical Etch Stop (for n-Type Silicon)....80
5.4.4 Etch Stop at Thin Films: Silicon on Insulator....81
5.5 High-Aspect-Ratio Micromachining....82
5.5.1 LIGA....83
5.5.2 Laser Micromachining....84
References....85

6. Microheaters for Gas Sensor....87
6.1 Introduction....87
6.2 Need of Microheater....88
6.3 Types of Microheater....89
6.3.1 Closed-Membrane Type....90
6.3.2 Suspended-Membrane Hotplates....91
6.4 Microheater Design Issues....93
6.5 Heater Material Selection....94
6.6 Heater Geometry Selection....95
6.7 Function of Interdigitated Electrode....98
6.8 Software Used....99
6.8.1 Temperature Distribution....100
6.8.2 Mechanical Stability....101
6.8.3 Thermal Response Time....102
6.9 Heating Power Consumption....103
6.10 Fabrication of Microheater....106
6.11 Microheater Array....108
References....110

Section II Sensor Applications

7. Semiconductors as Gas Sensors....119
7.1 Introduction....119
7.2 Development of Semiconductor Sensors....120
7.2.1 Fundamentals of Semiconductor Sensors....121
7.2.2 Classification of Semiconductor Sensors....121

7.2.3 Different Structures of Semiconductor Gas Sensors 122
7.2.3.1 Resistive-Type Metal Oxide–Based Gas Sensors 122
7.2.3.2 Schottky-Type Metal Oxide–Based Gas Sensor 122
7.2.3.3 Metal Oxide Homojunction Gas Sensor 124
7.2.3.4 Metal Oxide Heterojunction Gas Sensor 124
7.2.3.5 Mixed Metal Oxide Gas Sensors 124
7.2.3.6 MEMS Gas Sensors 125
7.3 What Is a Nanosensor? 126
7.3.1 Thin Film Sensors 127
7.3.2 Thick-Film Sensors 127
7.3.2.1 Thick-Film Materials 128
7.4 Solid-State Chemical Sensors 128
7.4.1 Metal Oxide Semiconductors 129
7.4.2 Nanocrystalline Metal Oxide Semiconductors 130
7.4.3 Adsorption of Oxygen: Analyses 130
7.4.4 Reaction between Gas (e.g. CH_4) and Oxygen 132
7.4.5 Role of Catalyst on Gas Sensing Mechanism 134
7.4.6 Thick- and Thin-Film Fabrication Process 135
7.4.6.1 Bulk Growth 135
7.4.6.2 Substrate Growth 137
7.4.6.3 Chemical Vapour Deposition 137
7.4.6.4 Sputtering 137
7.4.6.5 Chemical Route 138
7.4.7 Sensor Characterizations 138
7.4.7.1 X-Ray Diffraction 138
7.4.7.2 Determination of Crystal Size 139
7.4.7.3 Field Emission Scanning Electron Microscopy 139
7.4.7.4 Transmission Electron Microscopy 140
7.4.7.5 Photoluminescence Spectroscopy 140
7.4.7.6 Fourier Transform Infrared Spectroscopy 141
7.4.7.7 Qualitative Analysis 141
7.4.7.8 UV/VIS Spectroscopy 142
7.4.7.9 Raman Spectroscopy 142
7.4.8 Sensor Reliability Issues 143
References 143

8. Sensing with Graphene 147
8.1 Introduction 147
8.2 Properties of Graphene 150
8.2.1 Electronic Property 151
8.2.2 Mechanical Properties 152
8.2.3 Optical Properties 153

8.2.4 Electronic Transport ... 154
8.2.5 Anomalous Quantum Hall Effect ... 154
8.2.6 Magnetic Properties ... 155
8.2.7 Thermal Properties ... 156
8.3 Characterization Techniques ... 156
8.3.1 Optical Microscopy ... 157
8.3.2 Field Emission Scanning Electron Microscopy ... 157
8.3.3 Atomic Force Microscopy ... 158
8.3.4 Diffraction Imaging Electron Microscopy ... 158
8.3.5 Transmission Electron Microscopy ... 160
8.3.6 Raman Scattering ... 160
8.4 Synthesis of Single-Layer Graphene/Few-Layer Graphene ... 161
8.4.1 Micromechanical Exfoliation ... 161
8.4.2 Chemical Exfoliation ... 163
8.4.3 Epitaxial Growth on Silicon Carbide (SiC) ... 164
8.4.4 Chemical Vapour Deposition ... 165
8.5 Graphene Oxide ... 169
8.6 Potential Application ... 171
8.6.1 Graphene Sensors ... 171
8.7 Summary ... 174
References ... 175

9. Nanocrystalline ZnO-Based Microfabricated Chemical Sensor ... 185
9.1 Introduction ... 185
9.2 Device Structure: Vertical and Horizontal ... 186
9.3 Comparison of Vertical and Horizontal Structure ... 187
9.4 Metal–Insulator–Metal Structure ... 187
9.5 Nanocrystalline ZnO as Sensing Material ... 188
9.6 Sensing Layer Deposition by Chemical Route ... 189
9.6.1 Sol-Gel Method for Synthesis of ZnO Thin Films ... 189
9.6.2 Chemical Bath Deposition Technique ... 190
9.6.2.1 Growth Mechanism ... 191
9.6.3 Chemical Deposition Technique ... 191
References ... 192

10. Nanostructures for Volatile Organic Compound Detection ... 193
10.1 Introduction ... 193
10.2 Volatile Organic Compounds ... 193
10.3 Different Nanostructures ... 195
10.3.1 Fabrication Processes ... 195
10.3.2 Characterization Processes ... 196
10.4 Sensing Mechanism ... 196
10.5 Measurement Technique ... 197
10.6 Effect of Relative Humidity on VOC Detection ... 198
References ... 199

11. Sensor Interfaces 201
11.1 Signal Processing 201
11.2 Smart Sensors 202
11.2.1 System Components 205
11.3 Interface Systems 206
References 207

12. MEMS- and Nanotechnology-Enabled Sensor Applications 209
12.1 MEMS and Nanotechnology 209
12.2 Automotive Applications: An Elaborated Study 210
12.2.1 Safety 210
12.2.2 Vehicle Diagnostics/Monitoring 211
12.2.3 Engine/Drive Train 211
12.2.4 Comfort, Convenience and Security 212
12.3 Home Appliances 212
12.4 Aerospace 213
12.4.1 Turbulence Control 213
12.5 Environmental Monitoring 213
12.6 Process Engineering 214
12.7 Medical Diagnostic 215
References 216

Index 217

Preface

There is a significant interest in lowering the power consumption of semiconductor metal oxide–based gas sensors. Metal oxide gas sensors consuming low voltage and low power can be easily produced by combining micromachining and thin film technologies. The increasing demand for faster and more honest analysis has evolved the development of microelectromechanical systems (MEMS) for gas sensing and biosensing. MEMS are desirable for use in low-power investigative designs for their ability to employ and analyze small volumes.

The MEMS microheater is one of the key components of a chemical gas sensor. For semiconductor gas sensors, a uniform microheater temperature is a necessary requirement as it often enhances the operation of the sensor. The microheater that lies on top of the membrane should be maintained at a uniform temperature for maximum sensitivity. A uniform temperature implies minimization of the heater 'hot spot', which is a crucial requirement for heater reliability. Thus, the temperature uniformity depends mainly on the membrane materials and on the geometry of the microheater.

Low power consumption is a prerequisite for any type of sensor system to operate with acceptable battery lifetime. Power consumption of a MEMS-based gas sensor is found to depend mainly on thermal losses. It has been investigated that to achieve a very substantial reduction in power consumption, the dimension of the device has to be reduced, obviously in a cost-effective way. Micromachining of the silicon substrate is the only approach to obtain the desired geometries of the micromechanical structures.

Nanocrystalline oxide materials with high surface-to-volume ratios are gaining interest in the area of gas sensing because of their enhanced reaction possibilities between the adsorbed oxygen and target gases. The versatility of the sensing material affects their electrical, optical, and sensing applications. Among them, ZnO, owing to its unique features and advantages, attracted the attention of researchers worldwide to study its application in chemical sensors. Some of the unique features of ZnO include the availability of both n- and p-type materials, high electron mobility, wide band gap, compatibility with standard CMOS technology, and adequate lattice matching with Si and SiO_2 substrates. Graphene is another wonder material as it can sense an atomic level presence of toxic gases that would not be possible with metal oxides.

This book consists of 12 chapters. Section I consists of Chapters 1 through 6. The objective is to give a general concept about the microfabrication technology needed to fabricate a gas sensor on a MEMS platform.

Chapter 1 provides an introduction of MEMS for MEMS materials. A brief cleanroom concept is also given. The historical background of MEMS is also presented in this chapter.

Chapter 2 discusses the substrate materials used for MEMS. Contaminants affecting the substrate material are discussed. Some common insulating layers and their properties are elaborated.

Chapter 3 discusses the two types of deposition techniques, including the most popular one, chemical vapour deposition (CVD). Different types of CVD techniques are presented to provide a clear understanding to the reader. The methods of wire bonding are also discussed.

Chapter 4 illustrates the properties of photoresists, the types of photoresists and the photolithographic processes. Some advanced photolithographic techniques are also discussed in detail.

Chapter 5 deals with different micromachining techniques for the gas sensor platform, and bulk and surface micromachining. A comparison is made based on their requirements. A very specific etch stop technique is also discussed in this chapter.

Chapter 6 discusses the design issues of a microheater for MEMS-based sensors. In this chapter, a complete investigation is presented, including an electrothermal design of a microheater on a micromachined silicon platform, particularly applicable for relatively low-temperature (150°C–300°C) gas sensor applications. Microheater design issues along with their geometry are also presented. The modified structure along with the performance parameters, e.g. temperature distribution across the membrane and the temperature variation with power consumption, is critically discussed.

As this book is the amalgamation of MEMS with gas sensor, it is better to start with Chapter 7 in Section II.

Chapter 7 provides the synthesis technique of a nanocrystalline metal oxide layer which is to be deposited over the heater element. In this chapter, various characterizations have been furnished to identify the nanodimensional nature and some native defects of the thin film. Structural characterizations like XRD, FESEM and EDX confirming the crystal structure and morphology and determining the crystallite size are discussed lucidly. The impurity perspective has been analyzed by FTIR spectroscopy. An idea of Raman spectroscopy is also presented. Factors affecting gas sensing such as grain size, porosity of the material, nanosize effect and methane sensing mechanism are discussed thoroughly.

The investigation of graphene and its properties is currently a hot topic in physics, materials science and nanoscale science. Graphene is the strongest material ever measured, a replacement for silicon and the most conductive material ever discovered. It consists of a single layer of carbon atoms linked in a honeycomb-like arrangement. They behave well enough for use in nanoelectronics. Among the various potential applications of graphene, it can be an excellent sensor due to its 2D structure. Having a 2D structure, the entire volume is exposed to target gases, making it very efficient to detect adsorbed molecules. Gaseous molecules cannot be readily adsorbed onto graphene surfaces as graphene has no dangling bonds on its surface, so inherently graphene is insensitive. The sensitivity of graphene can be enhanced dramatically by adding some functional groups onto it, i.e. coating the film with polymers. The unsatisfied bond in the polymer layer acts as a concentrator which absorbs gaseous molecules. This absorption of molecules introduces a local change in the electrical resistance of the graphene layer, making it a good sensor.

Chapter 8 gives a detailed review about graphene; its different deposition techniques; and its important electronic, electrical and mechanical properties with its application as a gas sensor. Detection, alarm, and subsequent monitoring of poisonous and combustible gases (like H_2, CH_4) for application in domestic as well as industrial environment using a very-small-size, low-cost and low-power gas sensor is highly desirable.

Chapter 9 explores the different device structures that are possible in this fabrication process and their respective advantages. Then, zinc oxide is described in brief. Some low-cost, low-temperature synthesis techniques are introduced in the last section of the chapter.

Chapter 10 is concerned with volatile organic compound (VOC) detection. A few VOCs are mentioned. How relative humidity affects the sensing parameters is discussed.

In Chapter 11, different interface systems are discussed. A brief idea about smart sensors is given.

Finally, Chapter 12 presents the applications of MEMS and nanotechnology in different areas relevant to the sensor domain.

Authors

Sunipa Roy is assistant professor of electronics and telecommunications engineering at Guru Nanak Institute of Technology (JIS Group), Calcutta, India. She received her MTech in VLSI and microelectronics from West Bengal University of Technology in 2009 and her PhD in engineering from Jadavpur University in 2014. She has served as a senior research fellow of the Council of Scientific and Industrial Research, Government of India.

She is a member of the Institution of Engineers (India) and a member of the IEEE.

Her research interests include MEMS, nanotechnology, and graphene and its application as a gas sensor. She has published and presented several research papers in international journals and conferences and also supervises PhDs students.

Chandan Kumar Sarkar is professor of electronics and telecommunications engineering at Jadavpur University, Calcutta, India. He received his BSc (Hons) and MSc in physics from Aligarh Muslim University, Aligarh, India, in 1975 and earned his PhD from Calcutta University in 1979 and DPhil from the University of Oxford, Oxford, United Kingdom, in 1984. He has been teaching for more than 30 years.

In 1980, Dr. Sarkar received the British Royal Commission Fellowship to work at the University of Oxford. He worked as a visiting scientist at the Max Planck Laboratory in Stuttgart, Germany, as well as at Linkoping University in Sweden. Dr. Sarkar also taught in the Department of Physics at the University of Oxford and was a distinguished lecturer of the IEEE EDS.

Sarkar has served as a senior member of the IEEE and was chair of the IEEE EDS chapter, Calcutta Section, India. He served as fellow of the IETE, fellow of IE (India), fellow of WBAST, and member of the Institute of Physics, United Kingdom.

Dr. Sarkar's research interests include semiconductor materials and devices, VLSI devices, and nanotechnology. He has published and presented more than 300 research papers in international journals and conferences and also mentored 21 PhD students.

Section I

Fabrication Procedure

1

Introduction

1.1 Cleanroom Technology

To start with cleanroom technology, it is customary to understand the concept of cleanroom first. A cleanroom is where the intensity of bulk airborne particles can be controlled in a typical fashion. The controlling of the rate of airflow and its direction, temperature, relative humidity and pressure is required to have a cleanroom. Sub-micrometre-level airborne particles generally come from people, process, equipment, etc. A special filtering system is essential to have a cleanroom.

There are some strict protocols and methods in developing a cleanroom from its basic design level up to the construction phase. Obviously, a major user of cleanrooms is the semiconductor industry, but it is also used in electronics, pharmaceutical, biopharmaceutical and medical instrument industries, to name a few.

- *Class*: The number of particles present per cubic foot of air designates the class of cleanroom. To make it more clear, a Class 100 cleanroom can be defined as the room which is designed to never permit more than 100 particles (0.5 μm or larger) per cubic foot of air. It should also be noted that a typical office building air contains 500,000 to 1,000,000 particles (0.5 μm or larger) per cubic foot of air. Class 1,000 and Class 10,000 cleanrooms are designed to allow particles up to a maximum of 1,000 and 10,000, respectively.

  After manufacturing a cleanroom, it must be cleaned thoroughly to maintain the predetermined standards. There are several articles that elaborate the proper cleaning technique. Articles that discuss the method of training a professional cleaning staff are also available.

  There are several international standards used to measure the class, which restrict the definition of cleanroom. These standards actually define the level to which these particles have to be removed from the air. In Federal Standard 209E, the classes are taken from the maximum allowable number of particles that are 0.5 μm and larger per cubic foot of air.

TABLE 1.1
Cleanroom Standard

Air Particle (≥0.5 μm) Count per Cubic Foot	Federal Standard 209E	Air Particle (≥0.5 μm) Count per Cubic Meter	New ISO 14644-1 Class
100,000	Class 100,000	3,520,000	Class 8
10,000	Class 10,000	352,000	Class 7
1,000	Class 1,000	35,200	Class 6
100	Class 100	3,520	Class 5
10	Class 10	352	Class 4
1	Class 1	35	Class 3

The International Organization for Standards (ISO) is also developing a series of cleanroom standards. These cover a wide variety of important cleanroom issues such as design, testing, processes and defect. In ISO 14644-1, the classes are taken from the log (base 10) of the maximum allowable number of particles, which are 0.5 μm and larger per cubic metre of air (Table 1.1).

- *Various parts of cleanroom technology*: The following block diagram represents the cleanroom technology, which is divided into three broad areas. These areas can also be seen to parallel the use of the technology as the cleanroom user moves from first deciding to purchase a room to finally operating it [1].

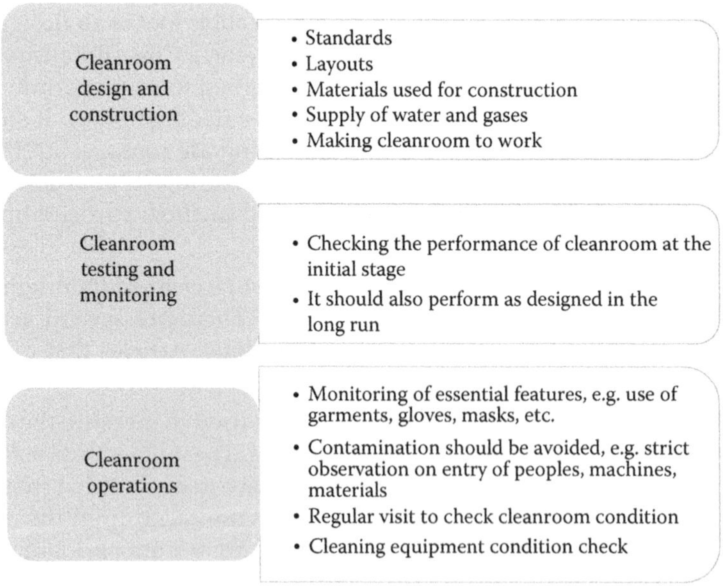

Cleanroom technology is based upon three criteria:

Cleanroom design and construction: It is essential to design and construct the room first. While designing, one must adopt the design standards, e.g. what type of design layout and construction materials can be used, and how requisite material would be supplied to the cleanroom.

Second, the most important aspect, is the continuous monitoring after the cleanroom has been installed and operational; it must be examined properly to confirm that it is working as determined. In the long run, it should conform to the standards that are required.

Finally, it is necessary to handle the cleanroom in a proper way to avoid contamination of the products. This includes proper checking of people's access, entry of materials and garments used, strictly following cleanroom rules and regulations and maintaining cleanliness of the room.

Contamination: Contamination is an unwanted phenomenon that causes surfaces of different substances to be exposed to contaminating particles. Contaminants are damaging in nature. Their effect is detrimental particularly in a miniature circuit. Contaminants (10 nm above) have severe effect in adhesion procedure. They reduce the stiction effect as a whole on a wafer or a chip. Generally, particles of 0.5 μm or larger create contamination. Salient features related to contamination are [2].

- *Contamination control*: This section will obviously give a better idea on the overall picture of contamination control. This topic should be given greater emphasis while treating contamination.
- *HEPA (high-efficiency particulate air filter)*: The efficiency should be atleast 99.97%, considering as low as 0.3 μm particle size; there are many other types of filtration mechanisms used to remove particles from gases and liquids. These filters are essential for providing effective contamination control.
- *Cleanroom architecture*: According to the rule, the air in a confined cleanroom area should move with uniform velocity along with parallel flow lines to attain airflow. This airflow is termed as a laminar flow. Any disorder can cause air turbulence, which can cause particle movement.
- *Electrostatic discharge (ESD)*: When two surfaces rub together, an electrical charge can be created. Moving air creates a charge. People touching surfaces or walking across the floor can create a triboelectric charge. Special care is taken in using ESD-protective materi als to prevent damage from ESD. Cleaning managers should work with their personnel to understand where these conditions may be present and how to prevent them.

- *Cleaning*: Cleaning is a fundamental process step in contamination control. For proper maintenance of cleanroom, applications and procedures need to be typed, focused and approved by the concerned authority. There are many issues associated with cleaning.

  Gloves, face masks and head covers are standard in every cleanroom environment. Jumpsuits are necessary to maintain a cleanroom. Physical behaviour like fast motion and horseplay can increase contamination. Psychological concerns like room temperature, humidity, claustrophobia, odours and workplace attitude are important.

1.2 Microelectromechanical System

Microelectromechanical system (MEMS) is basically a technology used to create tiny integrated devices that combine mechanical and electrical components. They are fabricated using batch processing techniques with size varying from a few micrometres to millimetres. The acronym of microelectromechanical system has originated in the United States, though in Europe, it is recognized as microsystem technology (MST), and in Japan, it is accepted as micromachining technology.

These devices (or systems) have the ability to sense, control and actuate on the micrometre scale and produce effects on the macroscale.

MEMS is interdisciplinary in nature, which has a wide and diverse range of technical areas including integrated circuit fabrication technology, mechanical engineering, materials science, electrical engineering, chemistry and chemical engineering, as well as fluid engineering, optics, instrumentation and packaging. MEMS can also be found in systems ranging across automotive, medical, electronic, communication and defence applications.

Moreover, the key aspect of MEMS device fabrication is that, in the technique, it is fabricated. The micromechanical components are fabricated by classy manipulations of silicon and other substrates using micromachining processes. Processes such as bulk and surface micromachining as well as high-aspect-ratio micromachining (HARM) selectively remove parts of the silicon or add structural layers to form the mechanical and electromechanical components. Integrated circuits are designed only to exploit the electrical properties of silicon; on the contrary, MEMS takes the advantage of other material properties like optical and mechanical. That is why these devices (or systems) have the ability to sense, control and actuate on the micrometre scale and generate effects on the macroscale.

MEMS consists of mechanical microstructures, microsensors, microactuators and microelectronics, all integrated onto the same silicon chip. This is shown schematically in Figure 1.1. MEMS components began appearing in numerous commercial products and applications including accelerometers

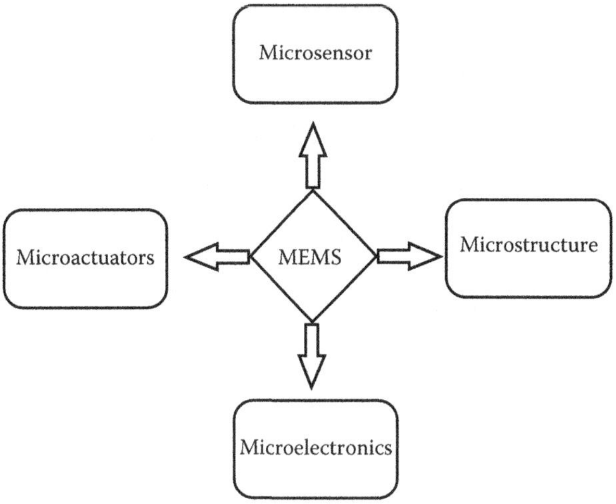

FIGURE 1.1
MEMS components.

used to control air bag deployment in vehicles, pressure sensors for medical applications and inkjet printer heads. Today, MEMS devices are also found in projection displays and in micro-positioners in data storage systems. However, the greatest potential for MEMS devices lies in new applications within telecommunication (optical and wireless), biomedical and process control areas.

1.2.1 History

1961: First silicon pressure sensor demonstrated.

1967: Invention of surface micromachining. Westinghouse creates the resonant gate field-effect transistor. Concept of surface micromachining with sacrificial layer first introduced.

1970: First silicon accelerometer demonstrated.

1979: First micromachined inkjet nozzle.

Early 1980s: First experiments in surface-micromachined silicon.

1982: Disposable blood pressure transducer.

1982: 'Silicon as a Mechanical Material'. Instrumental paper published to attract the scientific community – experimental data for etching of silicon first cited.

1982: LIGA (Lithography, Galvanoformung, Abformung) process demonstrated.

Late 1980s

Micromachining facilitates microelectronics industry, and extensive experimentation and documentation increases public interest.

1988: First MEMS conference organized.

Novel methods of micromachining modernized with an aim of improving sensors.

1992: MCNC (Microelectronics Center of North Carolina) starts the Multi-User MEMS Process (MUMPS).

1992: First micromachined hinge and beginning of the Bosch Deep Reactive Ion Etching (DRIE) process.

1993: First surface micromachined accelerometer sold (Analog Devices, ADXL50).

1994: DRIE is patented.

1995: BioMEMS rapidly develops.

Massive industrialization and commercialization.

2001: Triaxis accelerometers appear on the market.

2002: First nanoimprinting tools announced.

2003: For the purpose of volume applications, MEMS microphones introduced. Discera commences sampling of MEMS oscillators.

2004: Texas Instrument's digital light processing chip sales rose to nearly $900 million.

2005: Analog Devices embarked its 200 millions of MEMS-based inertial sensors.

2006: Packaged Triaxis accelerometers smaller than 10 mm^3 are becoming accessible. Dual-axis MEMS gyros appear on the market.

2006: Perpetuum launched vibration energy harvester.

1.2.2 Definitions and Classifications

This section defines some of the key terminologies and classifications associated with MEMS. This is introduced to help the readers including researchers become familiar with some of the more common terms of the field of micromachining. A more detailed glossary of terms is included in Figure 1.2.

MEMS is basically a method used to create miniaturized mechanical devices or systems. There are fundamental overlaps between different fields in terms of their fabrication technology and their application. Obviously, the classification of MEMS is extremely tricky with respect to domain or others. The classification of MEMS has been done judiciously, i.e. bulk

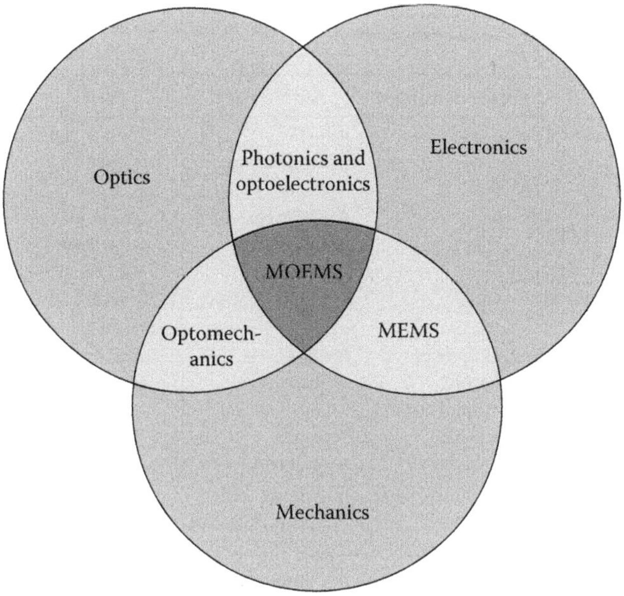

FIGURE 1.2
Classification of MEMS. (From Defense Advanced Research Projects Agency (DARPA), http://www.darpa.mil/MTO/.)

micromechanical part and surface micromechanical part. Bulk micromechanical part tends to use semiconductor processes to create a mechanical part. In contrast, a surface micromechanical part uses the deposition of a material on silicon, which does not constitute a MEMS product but one of the applications of microsystem technology.

Micromachining generally results in the creation of small parts such as lenses and cantilever with mechanical functionality. It is noteworthy that, in contrary to what its name suggests, MEMS may not always have mechanical parts; for example in MEMS-based gas sensor, bulk micromachining is done and is included in the MEMS family to minimize power loss without a mechanical part present. Micro-optoelectromechanical systems (MOEMS) are also a subset of MEMS and form the dedicated technology using combinations of miniaturized optics, electronics and mechanics. Such microsystems incorporate the use of microelectronics batch processing techniques for their design and fabrication. One portion of MEMS devices operating as actuators or sensors for biochemical processes and instrumentation is called BioMEMS.

1.2.3 Market and Application

Today, high-volume MEMS can be found in a variety of applications across the markets (Table 1.2).

TABLE 1.2
Applications of MEMS

Electronics	Communications	Automotive	Medical	Defence
Disk drives head	Fibre-optic network component	Internal navigation sensor	Blood pressure sensor	Munitions guidance
Inkjet printer heads	RF relays, switches and filters	Air-conditioning compressor sensor	Muscle stimulators and drug delivery systems	Surveillance
Projection screen televisions	Projection displays in portable communications devices and instrumentation	Brake force sensors and suspension control accelerometer	Implanted pressure sensors	Arming systems
Earthquake sensors	Voltage-controlled oscillators	Fuel level and vapour pressure sensors	Prosthetics	Embedded sensors
Avionics pressure sensor	Splitters and couplers	Air bag sensors	Miniature analytical instruments	Data storage
Mass data storage systems	Tuneable lasers	'Intelligent' tyres	Pacemakers	Aircraft control

Current MEMS devices include accelerometers for air bag sensors, inkjet printer heads, computer disk drive read/write heads, projection display chips, blood pressure sensors, optical switches, microvalves, biosensors and many other products that are all manufactured and shipped in high commercial volumes.

MEMS has been identified as one of the most promising technologies in the twenty-first century and has the potential to revolutionize both industrial and consumer products by combining silicon-based microelectronics with micromachining technology. Its techniques and microsystem-based devices have the potential to dramatically affect our lives and the way we live.

If semiconductor microfabrication was seen to be the first micromanufacturing revolution, MEMS is the second revolution.

It is not the aim of this chapter to detail all the current and potential applications within each market segment; the same has been dealt separately in Chapter 12.

1.2.4 Materials for MEMS

The requirement of optimal performance usually interprets the optimization of geometry, topology and mass of the structure. Material selection is really an exercise in design. Actually, the material selection depends on the availability and ease of fabrication using that material. In MEMS technology, popularity of silicon is mainly due to its abundant nature and ease of production and manipulation. However, several materials and processes have emerged in recent research in material science, which gives very good assurance while challenged with silicon. These new materials confer an opening to designers to evaluate these materials judiciously so that the performance can be at utmost level. This approach will thus offer two levels of optimization in device design. First, the best material should be selected (Table 1.3) and second, a proper framework should be chosen for evaluating the selected material. The materials chosen for the membrane of the microheater should combine low thermal conductivity (i.e. good dielectric) with high mechanical strength (i.e. large thickness). Some of the common MEMS materials are given in the following:

1. *Silicon (Si)*: Exceptional material that can be cost-effectively manufactured in single-crystal substrates. Each side is tetrahedrally coordinated with four other sides with other sublattices. Silicon is the material used to create most integrated circuits used in consumer electronics in the modern world. Just as silicon has dominated the integrated circuit industry, it is also predominant in MEMS. There are a number of reasons for this:
   a. Pure, cheap and well-characterized material already available
   b. Different types of matured and easily accessible processing technique
   c. The potential for integration with control and signal processing circuitry
2. *Silicon dioxide* (SiO_2): Most stable oxide used in the silicon-based IC industry, excellent thermal and electrical isolation. It can be easily deposited into the silicon wafer using comparatively low-cost deposition technique. There are three principal uses of silicon oxide in microsystems: (1) as a thermal and electrical insulator, (2) as a photolithographic mask during etching process and (3) as a sacrificial layer in surface micromachining. Silicon dioxide is a better etch resistant than silicon.
3. *Silicon nitride* (Si_3N_4): Silicon nitride (Si_3N_4) has many advantageous properties that are useful for MEMS. It renders an excellent barrier to diffusion of water and ions such as sodium. The ultra-high resistance to oxidation and many etchants (KOH, TMAH, etc.) makes it appropriate to be used as a mask for stronger etchant. The deposition process is expensive. Silicon nitride can be produced from silicon-containing gases and NH_3.
4. *Silicon carbide (SiC)*: The dimensional and chemical stability at very high temperatures (>600°C) makes the application of

TABLE 1.3
Some Common MEMS Materials

Materials	Thermal Expansion Coeff., 10^6/K	Thermal Conductivity at 300 K, W/M-K	Specific Heat, J/kg-K	Density, kg-M^3	Electrical Resistivity, Ω-m	Young's Modulus, GPa	Poisson's Ratio
Si	2.6–3.1	150	700	2,330	2300	125	0.27
SiO_2	0.5	1.4	1000	2,200	$(10^{12}–10^{14})$	75	0.17
Si_3N_4	3.3	30	1100	3,270	$(10^6 \times 10^{15})$	297	0.27
SiC	3.9	111	667	3,210	31.66×10^{-8}	390	0.15
Nickel	13	90.7	0.44×10^3	8,900	6.8×10^{-8}	200	0.31
Pt	8.8	70	130	21,090	10.6×10^{-8}	168	0.38
Au	14	295	130.2	19,300	2.2×10^{-8}	78	0.44
Air	3430	0.024	1.0005×10^3	1.239	1×10^{18}	1×10^{-8}	0.3
Poly silicon	2.8	29–34	702	2,330	3.22×10^{-7}	169	0.22
Porous silicon	2	1	850	466	Not specified	2.4	0.9

silicon carbide (SiC) in microelectronics industry possible. It is highly oxidation resistant even at very high temperatures. Silicon is found as a raw material in natural carbon-containing substances (coal, coke, wood chips, etc.); the heating of these materials at very high temperature in the electric arc furnace results in SiC precipitating to the bottom of the crucible. SiC film can be used as a passivation layer in micromachining for the underlying silicon substrate. The patterning of SiC can be easily done by shadow mask using dry etching technique.

5. *Polysilicon*: Polysilicon has become a principal material in surface micromachining. It is a good isotropic material in thermal and structural analyses. Silicon can be of three types: single crystal, polycrystalline and amorphous. Polycrystalline silicon is known as polysilicon. Properties of polysilicon vary with deposition conditions. Moreover, fabrication of MEMS devices using polysilicon require high annealing temperature (>900°C) and costly CVD technique. It exhibits unstable behaviour at greater than 250°C. It is not easy to deposit and etch. In heavily doped polysilicon, the resistivity is drastically reduced, which is effective to produce conductors and control switches.
6. *Silicon–germanium alloy (SiGe)*: Silicon alloyed with germanium has a wide range of applications; for MEMS applications, the layers of silicon–germanium alloy are grown using a graded buffer approach, resulting in a sophisticated micromachined structure [4,5].

   SiGe projects a platform for post-processing MEMS structures on top of prefabricated driving electronics; SiGe also owns good electrical and mechanical characteristics analogous to silicon. This alloy delivers electrical performance as per III–V semiconductors. SiGe has a current density of 10^6 A/cm^2 compared with 10^5 A/cm^2 for GaAs. The advantage of thin $Si_{1-x}Ge_x$ films is their basic compatibility with standard silicon technologies because a lattice mismatch of around 4% has been observed between pure Si and pure Ge, that is why such films are tetragonally distorted. Researchers are trying to explore its unique advantages, which will be revealed in the near future.
7. *Polymers*: Even though the electronics industry provides an economy of scale for the silicon industry, crystalline silicon is still a complex and relatively expensive material to produce. Polymers on the other hand can be produced in huge volumes, with a great variety of material characteristics. MEMS devices can be made from polymers by processes such as injection moulding, embossing or stereolithography, and are especially well suited to microfluidic applications such as disposable blood testing cartridges. It is also used as photoresists (SU8: epoxy-based photoresist can form layers up to 100 μm).

8. *Metals*: Metals can also be used to create MEMS elements. While metals do not have some of the advantages displayed by silicon in terms of mechanical properties, when used within their limitations, metals can exhibit high degrees of reliability. Metals can be deposited by electroplating, evaporation and sputtering processes. Commonly used metals include gold, nickel, aluminium, chromium, titanium, tungsten, platinum. and silver.

1.3 Significance of MEMS

> They tell me about electric motors that are the size of the nail on your small finger. And there is a device on the market, they tell me, by which you can write the Lord's Prayer on the head of a pin. But that's nothing; that's the most primitive, halting step in the direction I intend to discuss. It is a staggeringly small world that is below. In the year 2000, when they look back at this age, they will wonder why it was not until the year 1960 that anybody began seriously to move in this direction.
>
> **—Richard Feynman, "There's Plenty of Room at the Bottom" December 29, 1959, at the annual meeting of the American Physical Society at the California Institute of Technology**

MEMS has several distinct advantages as a manufacturing technology. In the first place, the interdisciplinary nature of MEMS technology and its micromachining techniques, as well as its diversity of applications, have resulted in an unprecedented range of devices and synergies across previously unrelated fields (e.g. biology and microelectronics). Second, MEMS with its batch fabrication techniques enables components and devices to be manufactured with increased performance and reliability, combined with the obvious advantages of reduced physical size, volume, weight and cost. Third, MEMS provides the basis for the manufacture of products that cannot be made by other methods. These factors make MEMS potentially a far more pervasive technology than integrated circuit microchips.

Though MEMS has awesome potential, the fabrication of MEMS devices is not an easy task. There are many technological challenges and constraints linked with the miniaturization of the devices that need to be addressed and overcome. MEMS technology finds applications in the following general domains:

Automotive domain

1. Air bag sensor
2. Vehicle security systems
3. Brake lights

4. Headlight levelling
5. Rollover detection
6. Automatic door locks
7. Active suspension

Consumer domain

1. Appliances
2. Sports training devices
3. Computer peripherals
4. Car and personal navigation devices
5. Active subwoofers

Industrial domain

1. Earthquake detection and gas shutoff
2. Machine health
3. Shock and tilt sensing
4. Toxic gas sensing
5. Optical device fabrication

Biotechnology

1. Polymerase chain reaction microsystems for DNA amplification and identification
2. Micromachined scanning tunnelling microscopes
3. Biochips for the detection of hazardous chemical and biological agents
4. Microsystems for high-throughput drug screening and selection
5. BioMEMS in medical and health-related technologies from lab-on-chip to biosensor and chemosensor

References

1. W. Whyte, *Cleanroom Technology: Fundamentals of Design, Testing and Operation*, John Wiley & Sons, Chichester, U.K., 2001.
2. R. Robinson, Removing contaminants from silicon wafers; To facilitate EUV optical characterization, BSc thesis, Brigham Young University, Provo, UT, August 2003.

3. Defense Advanced Research Projects Agency (DARPA), *An Introduction to MEMS (Micro-electromechanical Systems), Prime Faraday Technology Watch,* Wolfson School of Mechanical and Manufacturing Engineering, Loughborough University, U.K., January 2002, http://www.darpa.mil/MTO/.
4. E. A. Fitzgerald, K. C. Wu, M. Currie, N. Gerrish, D. Bruce and J. T. Borenstein, in *Microelectromechanical Structures for Materials Research* (Mater. Res. Soc. Proc. 518, Pittsburgh, PA, 1998), pp. 233–238.
5. J. T. Borenstein, N. D. Gerrish, M. T. Currie and E. A. Fitzgerald, in *Materials Science of Microelectromechanical Systems (MEMS) Devices* (Mater. Res. Soc. Proc. 546, Pittsburgh, PA, 1999), pp. 69–74.

2

Substrate for MEMS

2.1 Introduction

Microelectromechanical system (MEMS) is a triumph of twenty-first centuries and revolutionized the semiconductor industry by combining the microelectronics with micromachining technology. Though the term 'mechanical' is associated with MEMS, but in true sense, micromachined devices should not contain mechanical part always. But the purpose of MEMS technology is miniaturization that is needed to be addressed. Basically, MEMS is a platform on top of which some common microscopic mechanical parts, e.g. channels, holes, cantilevers, membranes, cavities and other structures, can be fabricated. Though micromachining is associated with MEMS, still the structures are not machined. Rather they are created using microfabrication technology suitable for batch processing for the integrated circuits.

The potentiality of MEMS fabrication is batch processing. Batch processing is highly economical; therefore, MEMS adds to economy in microelectronics industry.

The emphasis is given while choosing the MEMS for their mechanical properties than electrical. Although it depends on the particular application, mechanical properties have been encountered in MEMS such as high stiffness, high fracture strength, fracture toughness and high-temperature stability or chemical inertness.

The basic building block of MEMS devices is the substrate which is nothing but an object with macroscopic surface finish. In semiconductor electronics, the substrate is nothing but a slice of single crystal silicon which is commonly known as wafer. Wafer can also be made of other crystalline materials, e.g. quartz, alumina, GaAs and so on. These wafers should have a material quality, as high as possible, and at the same time, they should be cheaper to manufacture.

Among them the main advantage of using semiconductor (Si, Ge, GaAs) as a substrate material is that it can be used as a semiconductor as well as an insulator depending on the application in microelectronic industry. This flexibility has been achieved by the technique termed doping in which foreign materials are added to the semiconductor to convert them from semiconductor to the electrical conductor. The physics behind it is already elaborated in different books, so this is not included in this chapter.

From the technical point of view, the use of semiconductor as substrate material confers the integration of MEMS and CMOS structures into a truly monolithic device which is highly challenging; as all the processes such as etching, diffusion, deposition, etc., have already been established. However, much evolution continues to be made.

2.2 Silicon: The Base

2.2.1 Silicon as a Semiconductor

Single-crystal silicon is the principal material in the semiconductor industry. Other mostly used silicon compounds are SiO_2, SiC, Si_3N_4 and polysilicon. The polymers can also be used as substrate materials in MEMS and microsystems technology.

Among all, silicon is an ideal substrate material for MEMS. Silicon (Si) is the most abundant material on earth. Most of the time, it is found in compounds with other elements as a mixture.

Silicon is a popular semiconductor material and has already replaced other substrate materials due to the following reasons:

1. It is mechanically stable and is used in already advanced microfabrication technology.
2. It is lighter than Al and harder than steel. It has about the same Young's modulus as steel ($\sim 2 \times 10^5$ MPa), but is as light as aluminium with a density of about 2.3 g/cm^3.
3. Silicon is almost an ideal structural material. Miniaturized mechanical devices can be realized on silicon with high precision.
4. It has a melting point at 1400°C, which is about twice than that of aluminium. The high melting point makes silicon dimensionally stable even at high temperature.
5. Its thermal expansion coefficient is about 8 times smaller than that of steel and is more than 10 times smaller than that of aluminium.
6. Virtually, silicon has no mechanical hysteresis. It is thus an ideal candidate material for sensors and actuators.
7. Silicon wafers are superfinished, thus not suitable for coating. The extra thin-film layers should be integral structural parts of silicon, performing exact electromechanical functions.

8. The flexibility with silicon is much higher than other substrate materials.
9. The processing steps of silicon are already well established.

Silicon carbide (SiC), also known as carborundum, is a compound of silicon and carbon. Today, the application of SiC is mainly in high-temperature/high-voltage semiconductor electronics. SiC with high surface area can be produced from SiO_2 contained in plant material.

There is currently much interest in its use as a semiconductor material in electronics, where its high thermal conductivity, high electric field breakdown strength, and high maximum current density make it more promising than silicon in high-powered devices. SiC also has a very low coefficient of thermal expansion (4.0×10^{-6}/K) and experiences no phase transitions that would cause discontinuities in thermal expansion.

Other crystalline semiconductors including germanium (Ge) and gallium arsenide (GaAs) are also used as substrate materials due to analogous intrinsic features, but silicon is distinguished from other semiconductors for its property: it can be readily oxidized to form a chemically inert and electrically insulating surface layer of SiO_2 on exposure to oxygen and humidity.

Silicon is the material used to create most integrated circuits used in consumer electronics in the modern world. It is also an attractive material for the production of MEMS, as it displays many advantageous mechanical and chemical properties: single crystalline silicon is an almost perfect Hookean material. This means that when silicon is bent, there is virtually no hysteresis and hence almost no energy loss. This property makes it an ideal material, where many small motions and high reliability are needed, as silicon displays very little fatigue and can achieve service lifetimes in the range of billions to trillions of cycles.

Silicon is dominant as a substrate for MEMS, but research and development is ongoing with other non-semiconductor substrate materials including metals, glasses, quartz, crystalline insulators, ceramics and polymers. The ability to integrate circuitry directly onto the substrate is currently the underlying issue with today's MEMS substrate materials, hence the success of silicon.

2.2.2 Surface Contamination

Contamination is defined as a foreign material at the surface of the silicon wafer or within the bulk of the silicon wafer. Contamination can be of various types.

The main classes of contamination are

- Particles
- Metals

- Organics
- Volatile inorganic contamination
- Native oxide
- Microroughness

The big question that lingers is: does cleaning remove contamination or introduce it?

Semiconductor devices are very much prone to the contamination as the device size has been reduced significantly; surprisingly, the device itself has become comparable to the contaminant size. Contamination can be introduced in the wafer during different process steps, which uses chemicals. The environment is also responsible for the contamination on the wafer. Principally, contamination can be divided into three groups: ionic contamination, airborne molecular contamination (AMC) and particle contamination. Detailed discussion has already been published elsewhere [1]. In this chapter, a small description of different types of contaminants is given so that researchers may have some estimate before starting the device fabrication.

In contamination monitoring technique, particle-size monitoring is becoming a problem in advanced integrated circuits; in 130 nm processes, particles greater than 65 nm are monitored.

As the device scaling is ongoing, it may happen that monitored particle size will be identical to minimum line width, while wafer bonding particle is a major concern irrespective of line width.

2.2.2.1 Contaminants on Silicon Wafers

2.2.2.1.1 Metal Contaminant

It is not possible to avoid metal contamination; rather, it can be controlled by the steps of cleaning. Metal contamination on the surface can penetrate into the silicon bulk and can act as recombination centres for charge carriers. If this contamination precipitates at silicon/oxide interface or in the critical areas of the device, it can be problematic, because they affect diffusion profiles via their effect on crystal defects. Moreover, if metal contaminants get separated into the corresponding oxide during oxidation, they can prevent, retard or degrade oxide film growth and result in poor-quality oxides. Figure 2.1 shows different contaminants associated with silicon wafer.

Typical organic contaminants on silicon wafer are as follows:

- Phthalates, e.g. DEP, DBP, DOP
- Organophorus, e.g. TBP, TCEP, TEP, TPP, TCP
- Antioxidants, e.g. BHT, oxide of BHT

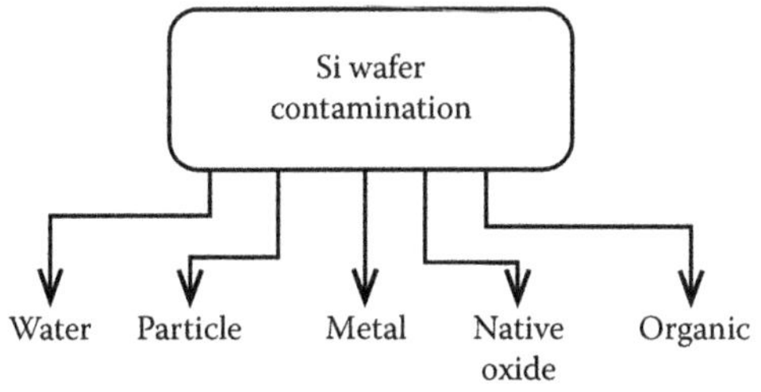

FIGURE 2.1
Different types of contaminants.

- Siloxanes, e.g. cyclo siloxanes (D3–D15)
- Adipates, e.g. DBA, DOA
- Amines
- Hydrocarbons

Effects of organic contamination on semiconductor devices are as follows:

- Degradation of gate oxide film
- Degradation of time-dependent dielectric
- Breakdown (TDDB) characteristics
- Haze degradation
- Poor oxide quality
- Silicon carbide formation
- Poor adhesion and conductivity
- Corrosion

Typical ionic contaminants on silicon wafer are as follows:

- Alkaline Na, K
- Transition Metals (Ni, Co, Fe, …)
- Dopants (Al, P, In, Ga, As, B, …)
- Acids (F^-, Cl^-,CH_3COO^-,Br^-, PO^{4--},SO^{4--})
- Bases (NH_3, amines)

Effects of ionic contamination on semiconductor devices

- Electrical instability: gate oxide leakage, retention
- Gate oxide integrity (GOI) degradation

- Shift of voltage threshold of the transistor device.
- Microfabrication process generated ionic contaminants can result pad corrosion, aluminium corrosion, defectivity on deep UV (DUV) and mid UV (MUV) resist. Air pollution can result salt deposition on lens, masks, and wafers.

2.2.3 Cleaning and Etching

2.2.3.1 Cleaning

Wafer cleaning is about contamination control, but it is also about leaving the surface in a known and controlled condition. This means damage removal, surface termination and prevention of unwanted adsorption. Therefore, many people prefer to call this activity surface preparation.

In the first step, possible organic or inorganic contaminations are removed from the wafer surface by a wet chemical treatment. The standard cleaning process developed by the company Radio Corporation of America (RCA) involves e.g. three cleaning steps based on solutions containing hydrogen peroxide.

In the following step, the wafer is heated to a temperature high enough to remove any remaining moisture from the wafer surface. This baking is done in order to improve the sticking of the photoresist layer.

To further enhance the sticking of the photoresist layer to the wafer surface, a liquid or gaseous 'adhesion promoter' (special chemical) is applied. Actually, the surface layer of SiO_2 on the wafer reacts with the chemical and produces a highly water repellent layer. It prevents the aqueous developer from penetrating between the photoresist layer and the wafer's surface and, by this, prevents the unwanted lifting of small photoresist structures in the pattern.

To start with any of the fabrication process, the wafer cleaning is required to remove the contamination, which occurs during the time of crystal growth and due to environment. The contaminations may be inorganic, organic or metallic types. To remove these contaminations from the wafer an easy, quick and cost-effective KAROW's cleaning method can be adopted.

2.2.3.2 Etching

Etching is used for removing patterns, healing of surface damage, cleaning the surface to remove contamination and fabricating 3D structures. Both wet chemical etching and dry etching are used.

- *Wet etch*: For normal etching of semiconductors, insulators and metal. Wet chemical etching is used extensively in semiconductor processing.

- *Dry etch*: Generally, plasma-assisted etching is used for high-fidelity pattern transfer. But the process is expensive.

Prior to any processing, semiconductor wafers are chemically cleaned to remove contamination that results from handling and storing. There are different chemicals and techniques used for different material etching.

2.2.3.2.1 *Wet Silicon Etching*

For Si, the most commonly used etchants are a mixture of nitric acid (HNO_3) and hydrofluoric acid (HF) in water or acetic acid (CH_3COOH).

Nitric acid oxidizes Si to form a SiO_2 layer. The oxidation reaction is as follows:

$$Si + 4HNO_3 \rightarrow SiO_2 + 2H_2O + 4NO_2$$

On the other hand, hydrofluoric acid is used to dissolve the SiO_2 layer. The reaction is as follows:

$$SiO_2 + 6HF \rightarrow H_2SiF_6 + 2H_2O$$

Water can be used as a diluent for HF. However, acetic acid is preferred, because it reduces the dissolution of nitric acid.

Some etchants dissolve a particular crystal plane of single-crystal Si much faster than another plane. This is an orientation-dependent etching. For a Si lattice, (111) plane has more available bonds per unit area than the (110) and (100) planes; therefore, the etch rate is expected to be slower for the (111) plane. A commonly used orientation-dependent etch for Si consists of a mixture of KOH mixed in water and isopropyl alcohol.

2.2.3.2.2 *Silicon Dioxide Etching*

The wet etching of SiO_2 is commonly achieved in a dilute solution of HF with or without the addition of ammonium fluoride (NH_4F). Adding NH_4F creates what is referred to as a buffered HF solution, also called buffered oxide etch (BOE). The addition of NH_4F in HF controls the pH level and replenishes the depletion of fluoride ions, thus maintaining stable etching performance. The etch rate of SiO_2 etching depends on etchant solution, etchant concentration, agitation and temperature. In addition, density, porosity, microstructure and the presence of impurities in the oxide influence the etch rate.

For example a high concentration of phosphorous in the oxide results in a rapid increase in the etch rate, and a loosely structured oxide formed by chemical vapour deposition (CVD) or sputtering exhibits a faster etch rate than thermally grown oxide.

SiO_2 can also be etched in vapour-phase HF. Vapour-phase HF oxide etch technology has a potential for submicron etching, because the process can be well controlled. Silicon nitride films are etchable at room temperature in HF or buffered HF solution and in a boiling H_3PO_4 solution. Selective etching of silicon nitride is done with 85% H_3PO_4 at 180°C because this solution attacks SiO_2 very slowly. The etch rate is typically 10 nm/min for Si_3N_4, but less than 1 nm/min for SiO_2. However, photoresist adhesion problems are encountered when etching nitride with boiling H_3PO_4 solution.

Better pattering can be achieved by depositing a thin oxide layer on top of the nitride film before resist coating. The resist pattern is transferred to the oxide layer, which then acts as a mask for subsequent nitride etching. Etching poly-Si is similar to that of crystalline Si. However, the etch rate is faster. Doping concentration and temperature may affect the etch rate.

2.2.3.2.3 Metal Etching

Al and its alloy films are generally etched in heated solution of phosphoric acid, nitric acid, acetic acid and DI water. The typical etchant is a solution of 73% H_3PO_4, 4% HNO_3, 3.5% CH_3COOH and 19.5% DI water at 30°C–80°C.

The wet etching of Al proceeds as follows:

$$HNO_3 \text{ oxidizes Al, and } H_3PO_4 \rightarrow \text{ then dissolves the oxidized Al}$$

Etch rate depends on etchant concentration, temperature, agitation of the wafers and impurities or alloys in the Al film.

2.2.3.2.4 Dry Etching

In wet chemical etching, reaction progresses equally in all directions. The major disadvantage in wet chemical etching is the undercutting of the layer underneath the mask, resulting in a loss of resolution in the etched pattern.

In pattern transfer operations, a resist pattern is defined by a lithographic process to serve as a mask for etching its underlying layer. Dry etching is anisotropic in nature and is highly directive. Most of the layer materials (e.g. SiO_2, Si_3N_4 and deposited metals) are amorphous or polycrystalline thin films. If they are etched in wet chemical etchant, the etch rate is generally isotropic (i.e. the lateral and vertical rates are the same).

If h_f is the thickness of the layer material and l is the lateral distance etched underneath the resist mask, we can define the degree of anisotropy (A_f) by

$$A_f \equiv 1 - \frac{1}{h_f} = 1 - \frac{R_l t}{R_v t} = 1 - \frac{R_l}{R_v}$$

where

t is the time

R_l and R_v are the lateral and vertical etch rates, respectively

For isotropic etching $R_l = R_v$ and $A_f = 0$

Basic steps of dry etching process:

1. Generation of etchant species
2. Diffusion to surface
3. Absorption
4. Reaction
5. Desorption into bulk

Dry etching methods include

- Plasma etching
- Reactive ion etching
- Sputter etching
- Vapour-phase etching

2.2.3.2.5 Plasma Etching

Fully/partially ionized gas is composed of equal number of positive and negative charges. High electric field initiates free electrons that collide with gas molecules and break the gas to ions. The electron concentration in the plasma is typically ~10^9 to 10^{12} cm^{-3}. At a pressure of 1 Torr, the concentration of gas molecules is 10^4–10^7 cm^{-3} times higher than that of the electron concentrations. This results in an average gas temperature in the range of 50°C–100°C. Thus, plasma-assisted dry etching is a low-temperature-driven process.

Plasma reactor technology in the IC industry has changed dramatically since the first application of plasma processing to photoresist stripping.

Dry etching differs from chemical etching in that dry etching has less etch selectivity to the underlying layer. Thus, the plasma reactor must be equipped with a monitor that indicates when the etching process is to be terminated (i.e. end point detection system). Laser interferometry of the wafer surface is used to continuously monitor etch rates and determine the end point. During etching, the intensity of laser reflected-off from a thin film surface, oscillates.

2.2.3.2.6 Reactive Ion Etching (RIE)

RIE has been extensively used in the microelectronics industry. A radio frequency capacitively coupled bottom electrode holds the wafer. This allows the grounded electrode to have a significantly larger area, because it is the chamber itself. The larger grounded area combined with the lower operating pressure (<500 mTorr) causes the wafer to be subject to a heavy bombardment of energetic ions from the plasma as a result of the large negative self-bias at the wafer surface. Selectivity of this system is relatively low because of strong physical sputtering.

2.2.3.2.7 *Sputter Etching*

Sputter etching is essentially RIE without reactive ions. The systems used are very similar in principle to sputtering deposition systems: substrate is now subjected to the ion bombardment rather than the target material used in sputter deposition.

2.2.3.2.8 *Vapour-Phase Etching*

In this process, the wafer to be etched is placed inside a vacuum chamber, in which gases are introduced. The material from the surface is dissolved in the gas ambient by the chemical reaction with the gas molecules. The applications of vapour-phase etching technologies are silicon dioxide etching using hydrogen fluoride (HF) and silicon etching using xenon difluoride (XeF_2). The etched structures are isotropic. The disadvantage of this technique is that the production of by-products arises from the chemical reaction, which may be due to the condensation of the intermediate compound on the surface of the substrate.

2.2.3.2.9 *Dielectric Etching*

Because of higher bonding energies, aggressive ion-enhanced, fluorine-based plasma chemistry is required. The normally used compounds are as follows: CF_4, CHF_3 and C_4F_8. High ion-bombardment energies are required to remove the polymers from the oxide.

2.2.3.2.10 *Interconnect Metal Etching*

Al, Cu and W are the mostly used metals for interconnections. Chlorine-based chemistry (e.g. Cl_2/BCl_3 mixture) is widely used for Al. Cu halides are non-volatile in nature, and thus, plasma etching is not suitable. The damascene process is used to form Cu interconnection instead of dry etching. In damascene process, first a trench/canal on the dielectric layer is created by dry etching and then it is filled with metal.

In dual-damascene process, a second level is involved where a series of holes are (contacts or vias) are etched and filled in addition to the trench.

2.3 Dielectrics

Semiconductor technologies are advancing day by day and there is a requirement of new materials always. Permuting the materials and their properties at the atomic scale has become very essential. Thousands of dielectric films are on the way to be used in optics and semiconductors. Most common dielectrics discussed in this chapter are as follows:

- *SiO_2*: This is the only native oxide of a common semiconductor which is stable in water and at elevated temperatures, an excellent electrical insulator, a mask to common diffusing species and capable of forming a nearly perfect electrical interface with its substrate. It is non-absorbing ($k=0$) over most wavelengths and is usually very close to being stoichiometric (i.e. the Si:O ratio is very close to exactly 1:2). Thermally grown SiO_2 is particularly well behaved and is commonly used for thickness and refractive index standards. The Filmetrics Spectrometer systems can measure thickness of the SiO_2 layer ranging from 3 nm to 1 mm thick.
- *Si_3N_4*: The measurement of silicon nitride films is more challenging than that of many other dielectrics, because the film's Si:N ratio is rarely exactly 3:4, and thus, the refractive index must usually be measured along with the film thickness. To complicate matters, oxygen is often unintentionally incorporated into the films, creating silicon oxynitride to some degree.

2.3.1 Silicon Dioxide (SiO_2)

Depending on the application, the dielectric layer can be grouped into three categories:

1. The active dielectric layer that plays an active role in device operation.
2. The layer that is required during device processing such as silicon nitride and silicon carbide as antireflection coating (ARC) layer needed in microelectronics lithographic step to lessen the image distortion resulted from the reflection of light during lithography or it can be used as a etch stop layer during wet etching.
3. The insulating layer called intermetallic dielectric is needed to isolate two metallic layer and two metal lines. Similarly, when isolation of the device from the surrounding is needed, the insulating layer behaves as a passivation cap.

The growth of silicon dioxide is most crucial in the fabrication of MOS transistors [2]. The attributes of SiO_2 which make it appealing for the semiconductor industry are as follows:

- It is easily deposited on various materials and grown thermally on silicon wafers.
- It is resistant to many chemicals used during the etching of other materials, while allow itself to be selectively etched with certain chemicals or dry-etched with plasmas.

TABLE 2.1
Mechanical Properties of Few Substrate Material

Materials	Young's Modulus, GPa	Dielectric Constant (e_r)	Yield Strength, GPa	Knoop Hardness, kg/mm²
Si	150	11.8	7	950
SiO_2	73	3.9	8.4	820
Alumina	50.2	9.82	21.5	1370
Quartz (Silica)	10.6	3.78	4.5	741
SiC	700	3.2	21	2480
TiC	497	20	4.9	2470
Si_3N_4	385	3.1	14	3486
Iron	186	7.8	12.6	400
Al	70	2.7	0.17	130
Stainless steel	200	7.9	2.1	660

- It can be used as a blocking material for ion implantation or diffusion of many unwanted impurities.
- The interface between silicon and silicon dioxide has relatively few mechanical and electrical defects, although with newer technology nodes and reduced geometries, even slight defects must be addressed.
- It has a high dielectric strength and a relatively wide band gap, making it an excellent insulator.
- It has a high-temperature stability of up to 1600°C, making it a useful material for process and device integration.

Selected properties of SiO_2 films can be referred from Table 2.1.

2.3.2 Silicon Nitride (Si_3N_4)

Silicon nitride (Si_3N_4) is produced by the chemical reaction:

$$3SiCl_2H_2 + 4NH_3 \rightarrow Si_3N_4 + 6HCL + 6H_2$$

Si_3N_4 is used as an excellent barrier to diffusion of water and ions. Its ultra-strong resistance to oxidation and many etchants make it a superior material for masks in deep etching. It is also used as high mechanical strength electric insulators. Selected properties of Si_3N_4 films can be referred from Table 2.1.

High-k gate dielectrics is a new term in modern microelectronics industry and can be used as alternative gate dielectrics for the 65 nm CMOS technology and beyond to replace conventional SiO_2 or silicon oxynitrides (SiO_xN_y) due to its high dielectric constant and large band gap.

2.3.3 Low-Temperature Oxidation

It can be categorized as

- High-pressure oxidation
- Plasma oxidation
- Rapid thermal oxidation (RTO)

Oxidation at high pressure produces a substantial acceleration in the growth rate. Thermal oxide layers can therefore be grown at low temperature in run times comparable to typical high temperature in order to reduce dopant diffusion and suppress oxidation-induced defects.

Anodic plasma oxidation has all the advantages associated with the high-pressure technique and also offers the possibility of growing high-quality oxides at even lower temperatures. Plasma oxidation is a low-pressure process usually carried out in a pure oxygen discharge. The plasma is sustained by either a high-frequency or DC discharge. Placing the wafer in the uniform density region of the plasma and biasing it slightly negatively against the plasma potential allow it to collect active charged oxygen species. The oxidation rate typically increases with higher substrate temperature, plasma density and substrate dopant concentration.

RTO is increasingly used in the growth of thin, high-quality dielectric layers [3]. The primary issues that differentiate RTO from conventional thermal oxidation are the more complex chamber design, radiation source, as well as temperature monitoring. From the point of view of oxide-growth kinetics, RTO may be influenced by both thermally activated processes and a non-thermal, photon-induced process involving monatomic O atoms generated by UV and creating a parallel oxidation reaction that dominates at lower temperature. RTO growth kinetics exhibits activation energies differing from those measured in conventionally grown oxides.

2.3.4 Oxide Properties

Many studies on silicon dioxide actually deal with the properties of its point defects, which can be both native and radiation induced, since they can significantly affect the macroscopic properties of the material. Impact makes an atom of the lattice move from its site to an interstitial position leaving a vacancy [4], significantly enhancing the damage, though thermal treatment can be used to delete radiation effects. Heating an irradiated sample above specific threshold temperatures determines a change in the concentrations of point defects.

In fact, point defects can act as charge traps, or change the refractive index, or give rise to optical absorption (OA) and photoluminescence (PL) activities because of the presence of their energy levels in the band gap. Particular attention has always been devoted to the study of the properties of point defects in silica, due to the fact that their presence is often associated with OA and PL activities, and can compromise the characteristic features of the material.

OH groups are one of the main impurities in silica. Great efforts were made to improve the production techniques, aiming to obtain OH-free materials. The generation of OH groups was also observed after thermal treatments in hydrogen, instead of water, atmosphere at temperatures higher than 400°C [5,6].

If a charge is present close to the Si/SiO_2 interface, it can induce a charge of the opposite polarity in the underlying silicon, thereby affecting the ideal characteristics of the device, such as the threshold voltage of an MOS capacitor. These charges originate from structural defects related to the oxidation process, metallic impurities, and bond-breaking processes. A low-temperature hydrogen anneal at 450°C effectively neutralizes most interface-trapped charges. The interface with silicon always results in electronic trap levels and some negative interface charge. Typical interface defect density is $\approx 10^{11}$ cm^{-2}.

The quality of the oxide depends primarily on the density. Since there is some flexibility in how closely the Si-O tetrahedral can be packed, there will be variations in density. Increased density manifests itself in terms of higher breakdown fields (important in the gate layer of a MOSFET) and lowered volatility (harder to react with other materials, slower etch rates). If oxide quality is a concern, we need to make the growth conditions conducive to higher density. In general, this means slower growth rates and higher temperatures. Thus, we expect dry oxidation to give better oxide than wet. It is also possible to 'anneal' an oxide to increase the density. For example leave it in the furnace at an elevated temperature for some time. CVD oxides are grown at much lower temperatures and higher growth rates, leading to much lower quality than thermally grown oxides.

References

1. A. Shiue, S.C. Hu, C.H. Lin and S.I. Lin, Quantitative techniques for measuring cleanroom wipers with respect to airborne molecular contamination, *Aerosol Air Qual. Res.*, 11, 460–465 (2011).
2. O. Biserica, P. Godignon, X. Jorda, J. Montserrat, N. Mestres and S. Hidalgo, Study of AlN/SiO_2 as dielectric layer for SiC MOS structures, *CAS 2000 Proceedings, International*, Sinaia, Romania, 10–14 October, 2000, vol. 1, p. 205.
3. C.E. Weintraub, E. Vogel, J.R. Hauser and Y. Nian, Study of low-frequency charge pumping on thin stacked dielectrics, *IEEE Trans. Electron Devices*, 48, 2754–2762 (2001).
4. A. Dunlop, F. Rullier-Albenque, C. Jaouen, C. Templier and J. Davenas, eds. *Materials under Irradiation*. Trans Tech Publications, Switzerland (1993).
5. K. Kajihara, M. Hirano, L. Skuja and H. Hosono, Luminescence of non-bridging oxygen hole centers in crystalline SiO_2, *J. Appl. Phys.*, 98, 043515 (2005).
6. J.E. Shelby, Reaction of hydrogen with hydroxyl free vitreous silica, *J. Appl. Phys.*, 51, 2589–2593 (1980).

3

Deposition

Deposition is one of the main fabrication tools used in microelectromechanical system (MEMS) fabrication to deposit thin films of materials. Deposition actually changes the surface properties of the base material on which it is deposited. In this chapter, we discuss the deposition of thin-film thickness between a few nanometres and about 100 μm. In MEMS technology, the film is patterned and can be subsequently etched away using the steps elaborated in the lithography chapter of this book.

MEMS deposition technology can be categorized into two ways. The first one is responsible for chemical reaction, i.e. chemical vapour deposition (CVD), electrodeposition, epitaxial growth, etc.

The second deposition process is due to physical reaction, i.e. physical vapour deposition (PVD). It is not possible to cover the whole thing in a single chapter; still, the major parts are covered to give the reader an idea of the overall concept.

3.1 Physical Vapour Deposition

PVD is basically applied to deposit thin film of different materials onto a substrate. The substrate may be a glass or alumina or silicon. The most common methods of PVD of metals are thermal evaporation, e-beam evaporation, plasma deposition and sputtering. The structure of the thin film can be tailored by varying the deposition technique. Metals and different metal compounds can be deposited by PVD.

Evaporation is one of the easiest routes of PVD. In this process, the source material is heated above its melting point in a vacuum chamber. The evaporated atoms pass with a high velocity from source to target and follow straight-line trajectories. The sources can be melted by various methods. One is by resistive heating or radio frequency (RF) heating, or through a focused electron beam. Among all, thermal and e-beam evaporations are extensively used till date due to their simplicity in nature. But presently, sputtering technique is used in modern ULSI (ultra-large scale integration: more than one million components per chip) circuit.

Atomic layer deposition (ALD) is another important PVD technique and is mainly used for high-aspect-ratio structures. The deposition of materials is reduced to one monolayer at a time. After successive deposition, an extremely conformal, defect-free layer is formed on the high-aspect-ratio structures. In this technique, precursor reacts with the surface forming only a monolayer of the material. The precursor is injected into the chamber, and then the surface is again hydroxylated with water vapour or oxygen followed by another purge. These two steps are then repeated until the desired thickness of the material is achieved. ALD has a vast array of applications from semiconductors, MEMS, nanostructures and optics to wear-resistant coatings.

A brief study of the different thin-film deposition techniques is explained in the following.

3.1.1 Vacuum Technology for MEMS

Deposition and vacuum, these two terms, are closely related because most of the deposition process occurs in vacuum condition. It is the only way to achieve the required quality. Particularly, for thin-film deposition, the study of vacuum technology is extremely important and has been used since the last two decades. Several authors have worked out on vacuum technology [1], and being a very specific topic of research, the detailed discussion is not included here; this section will be restricted to a discussion of applications in the important fields of coating technology.

In industrial vacuum processes, four pressure regions can be distinguished:

- Rough vacuum 10^5–10^2 Pa
- Medium vacuum 10^2–10^{-1} Pa
- High vacuum 10^{-1}–10^{-5} Pa
- Ultra-high vacuum 10^{-5}–10^{-8} Pa

Right now, the fields of vacuum technology are manifold:

- Electronics (e.g. cathode ray tube [CRT], drying of electric components)
- Deposition techniques (e.g. evaporation, sputtering)
- Food and pharmaceutical industry
- Chemical analysis techniques (e.g. Rutherford backscattering spectrometry [RBS], scanning electron microscope [SEM], secondary ion mass spectrometry [SIMS], atomic emission spectroscopy [AES])

There are many advantages of using vacuum. A very good quality of surface morphology of the deposited film has been observed while using vacuum processes. The properties of deposited thin films can be measured by in situ monitoring technique. There is a high rate of repeatability, which is

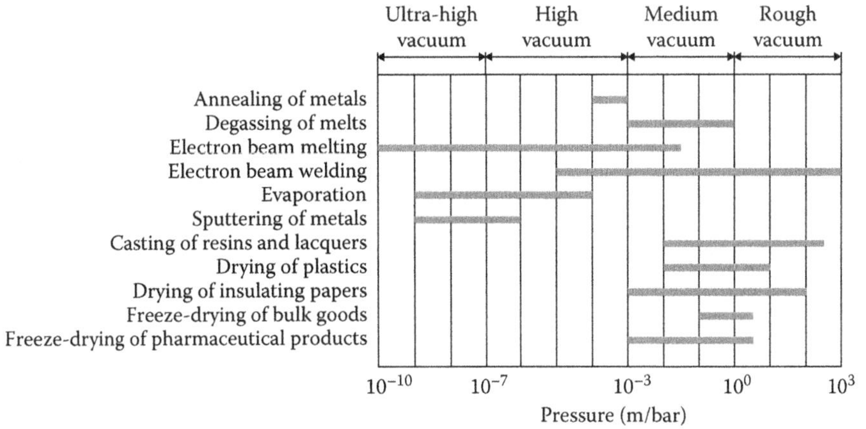

FIGURE 3.1
Bar chart showing vacuum regions.

not possible with other deposition techniques. But the controllability of the process is poor; in that sense, aqueous chemical root is better.

The advantages associated with this process are as follows: it is economical as the material consumption is minimal. The process throughput is very high; that is why this technology is much suitable for batch production. Less exposure to unsafe gases and chemicals makes it environmentally friendly. There are no waste products in most of the cases (Figure 3.1).

In all vacuum coating methods, layers are formed by the deposition of materials from the gas phase. The coating material may be formed by physical processes such as evaporation and sputtering or by chemical reaction.

The problem associated with chemical deposition technique of the film is erratic, and the thickness is non-uniform. The main advantage of vacuum is the formation of the conformal coating ranging from several nanometres to more than 100 mm with high quality and superb repeatability of coating property. Any type of substrate material (metals, glass, alloys, plastics, ceramics, etc.) can be used as this is not directly in contact with heat. Sometimes substrate heating is obviously required, but this may vary with deposition condition. In addition to metal and alloy coatings, layer-by-layer structure in sandwich form can also be deposited using different deposition techniques, which is mentioned later. Various chemical compounds have been used as precursors, and it is applied one at a time. A major benefit of using vacuum coating over other methods is that the growth rate can be controlled accurately as there is a provision of in situ characterization; many required coating properties can be manipulated, such as structure, thickness, electrical conductivity or refractive index by applying a specific coating method and varying the process parameters such as temperature and partial pressure for a certain coating material.

3.1.2 e-Beam Evaporation

Electron beam evaporation is a very important tool of PVD method for depositing thin films of metals, oxides and semiconductors in a high-vacuum environment. Ultra-high-purity coating material is placed inside a vacuum chamber, typically as pellets in a crucible. Electron energy is used to heat these pellets, causing the coating material to enter the gas phase. Due to the vacuum environment, the evaporated particles can travel to the substrate without colliding with foreign particles. They then condense on the substrate surface in a thin film. The schematic of the set-up is shown in Figure 3.2.

Electron beam evaporation is used to deposit electronic and optical films for the semiconductor industry and has applications in displays and photovoltaics. High-melting-point materials can be deposited at high deposition rates, making this a preferred process for refractory metal and ceramic films. Coating thickness is from 2 nm up to 200 nm (thickness >200 nm by approval). Currently available in the materials library are metals (Al, Cr, Ti, Au, Ag), oxides (SiO_2, TiO_2, Al_2O_3, ITO), fluorides (MgF_2) and semiconductors (Si, Ge).

3.1.3 Thermal Evaporation

Resistive thermal evaporation is another form of PVD and can be used to deposit metals and organic and inorganic polymers. In this method, electrical energy is used to heat a filament, which in turn heats the deposition material to the point of evaporation. The process can be performed at very high levels of vacuum to eliminate collisions with foreign particles and therefore fewer tendencies to introduce film impurities. High deposition

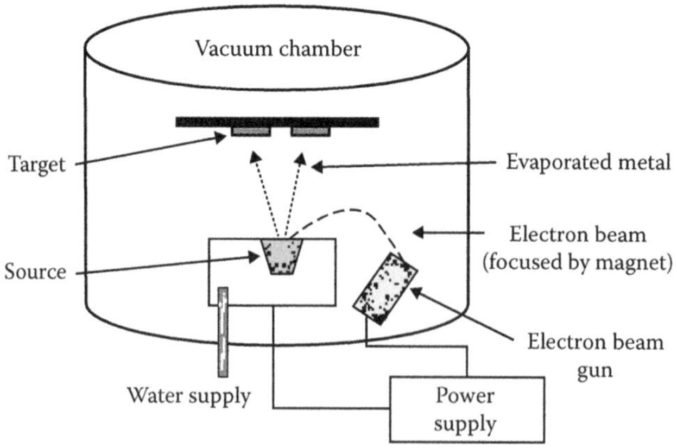

FIGURE 3.2
e-Beam evaporation unit.

rates can be achieved, and lower-energy particles can reduce substrate damage.

Sequential as well as co-deposition of thin films is possible with thermal evaporation technique. In the evaporation process, the material to be deposited is heated to the melting point of that material to generate sufficiently high vapour pressure, and the desired evaporation or condensation time is set. The simplest sources used in this evaporation are wire filaments of higher melting point, boats of sheet metal or ceramics, which are heated by passing a temporary flow of electrical current through them.

There are also some restrictions in the process:

- The filament temperature should be increased gradually; otherwise, the desired thickness will not be achieved, and the growth rate will be non-uniform.
- The source material should be consumed fully at the time of evaporation.
- If the source holder is evaporated at higher temperature for a longer time, the contamination may occur.
- In addition, the chemical reactions between the holder and the source material to be evaporated can also cause contamination.
- Continuous heating and cooling of the evaporator destroy the evaporator in the long run. Proper replacement is mandatory.

The schematic of the set-up is shown in Figure 3.3.

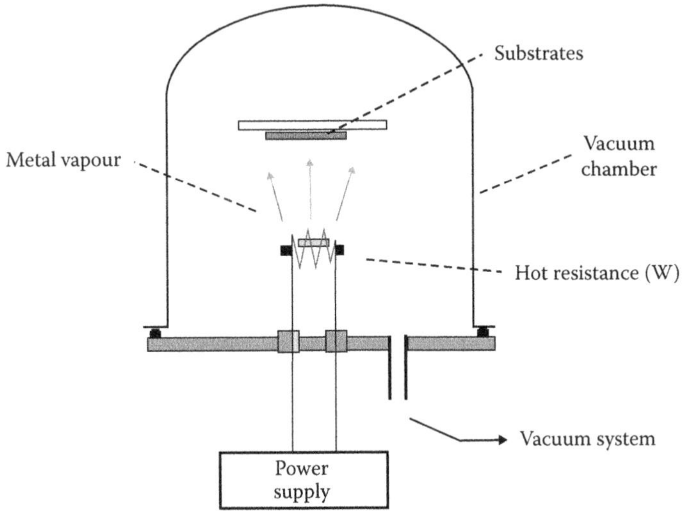

FIGURE 3.3
Thermal evaporation unit.

3.1.4 Sputtering

Sputtering is a PVD method that involves deposition of thin films in a vacuum environment. During this process, a solid material and a substrate are positioned separately within a vacuum system. A high-energy argon ion plasma stream is targeted at the material, resulting in the subject material being ejected and deposited onto the substrate, creating a thin film.

As this is not an evaporative process, the temperatures required for sputtering are lower than evaporation methods. This makes it one of the most flexible deposition processes, and it is particularly useful for depositing materials with a high melting point or a mixture of materials, as compounds that may evaporate at different rates can be sputtered at the same rate. Certain processes will benefit from improved film adhesion due to higher impact energy.

The sputtering process is used extensively in the semiconductor industry, screen displays, photovoltaics and magnetic data storage. Sputtering can be used to deposit a wide variety of thin films including metals, oxides, nitrides and alloys. The schematic of the set-up is shown in Figure 3.4.

Positive ions in the plasma are accelerated towards the target. If sufficiently energetic, the collision will cause one (or more) atom from the target to be knocked loose (sputtered off). The sputtered atoms will diffuse through the plasma region. Some will impinge on the substrate. If the sputtered atoms have a high sticking coefficient (low vapour pressure), they will adsorb the wafers. The sputtered atoms have enough energy so much so that there will be some surface mobility, allowing the adsorbed atoms to move around.

In ion beam sputtering, ions are accelerated towards the target and impinge on its surface. The sputtered material deposits on a wafer that is placed facing the target. The ion current and energy can be independently adjusted. Since the target and the wafer are placed in a chamber that has

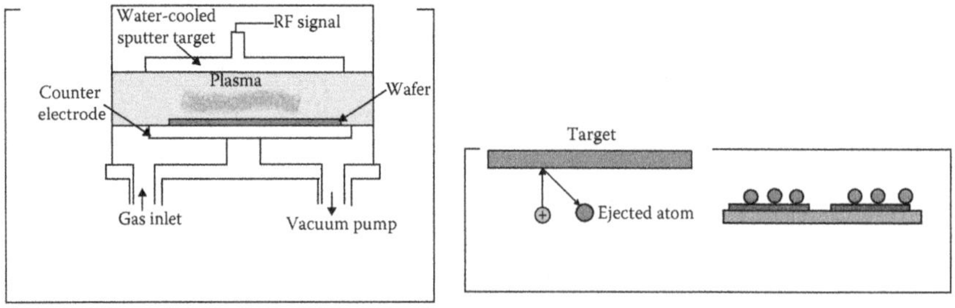

FIGURE 3.4
Sputtering unit with sputtered atom.

lower pressure, more target material and less contaminations are transferred to the wafer.

3.1.5 Molecular Beam Epitaxy

Molecular beam epitaxy (MBE) is a PVD process. Unlike LPE and VPE, there is no vapour-phase deposition process. Its advantages over other epitaxial processes are the extreme precise control of doping and the possibility of heterostructure formation. It involves the reaction of more than one beam of molecules maintained in ultra-high vacuum condition (10^{-8} Pa). Because of this ultra-high vacuum condition, the growth rate is slower (1 μm takes several hours) than other epitaxial deposition techniques; that is why for industrial application, MBE technique is not suitable. But one can precisely control the chemical composition or doping profile of the device. A single-crystal multilayer structure down to subatomic level is only possible with MBE, which means that MBE is the ultimate in film deposition control, purity and in situ characterization. The surface morphology is abrupt when there is a problem with chemical composition or lattice matching. The deposition process is totally different from VPE. The main technique is evaporation. Elemental materials are kept in the effusion cells from where the individual molecules flow through a hole/orifice without collision between the molecules. Vapours from the effusion cell follow a straight line, and the mean free path of molecules is very large, which prevents the chances of collision between the molecules. If there is any type of turbulence in the path of the beam, then the growth of deposition is hampered. The effusion cell is sometimes called Knudsen cell. The Knudsen cell is basically a chamber where solid materials are evaporated at high temperature as melting points of these solids are high. The temperature control is critical (~1600°C). The substrate holder is also kept at certain temperature (~400°C–800°C) where the substrate is kept. The vapour is mixed with certain stoichiometric ratio. If temperature somehow changes, then the partial pressure of precursor molecule also changes, which will degrade the growth rate. So, precise control of temperature is essential in the MBE process. Each orifice is operated with a shutter, which must be interrupted to control the composition and doping of the film. A number of effusion cells can be arranged to deposit onto the substrate. Each cell contains different materials, say for an n-type layer, phosphorus-containing shutter is opened; then, for the p-type, boron shutter is opened. In this way, a layer-by-layer structure (e.g. npn transistor) can be grown. The substrate is placed at a certain distance. Chilled water is used to make the chamber cold. Only effusion cell and substrate holders are heated. The set-up is shown in Figure 3.5. Lots of pumps are connected with the chamber to supply cold water. The machine is quite complicated. This epitaxial technique mainly comes out for the formation of GaAs/AlGaAs system, which is required for the first window of fibre-optic communication.

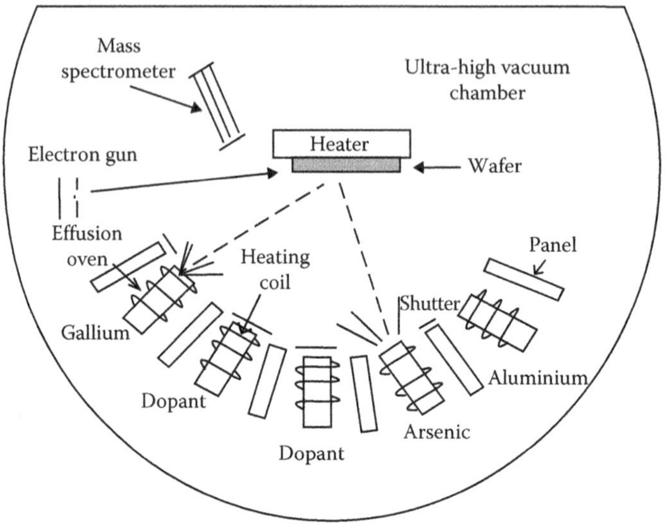

FIGURE 3.5
Set-up of molecular beam epitaxy.

3.2 Chemical Vapour Deposition

CVD is attractive for metallization because it offers coatings that are conformal, has good step coverage and can coat a large number of wafers at a time. The basic CVD set-up is the same as that used for the deposition of dielectrics and polysilicon. Low-pressure CVD is capable of producing conformal step coverage over a wide range of topographical profiles, often with lower electrical resistivity than that from PVD.

3.2.1 Atmospheric Pressure CVD (APCVD)

Epitaxial films are single-layer film. These films should be of very good quality so that it can be used in VLSI Microelectronics application. The films are polycrystalline or amorphous in nature. This type of film can be deposited by the CVD method. It is highly used in industrial processes and offers low cost and higher throughput.

CVD is defined as the formation of non-volatile solid films on a substrate by the reaction of vapour-phase chemicals that contain the liquid. These vapour-phase chemicals (gaseous or liquid phased), are vaporized and flown into the reaction chamber with partial pressure for the reaction to take place. The set-up is shown in Figure 3.6.

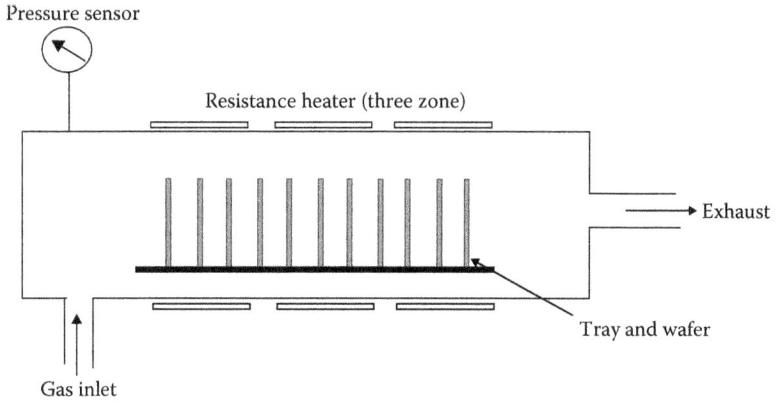

FIGURE 3.6
Set-up of chemical vapour deposition at atmospheric pressure.

Basically, CVD reaction follows these steps:

- Transportation of reactant gases on the substrate surface.
- The flow of gases is well controlled. The gas may be a source gas or a carrier gas. For Si_3N_4 deposition, NH_3 or H_2O is the source gas and H_2 is the carrier gas.
- Once reactant gases are flown, then adsorption or chemisorption occurs on the substrate surface.
- Heterogeneous reaction catalysed by the substrate surface.
- Desorption of gaseous by-product that moves away from the substrate surface.

3.2.2 Plasma CVD

Plasma-enhanced CVD (PECVD) is a special technique of CVD process where glow-discharge plasma persists in a reaction chamber. This technology was developed first to satisfy the need of the semiconductor industry while depositing a passivation layer of silicon nitride with a low-temperature deposition technique. Deposition temperatures range from room temperature to 400°C. This method is particularly useful where the film could not be exposed to higher temperatures like 1000°C, which is a standard in CVD.

The reason behind this reduced temperature of operation is that the thermodynamics of surface reactions changes in the presence of plasma and considerably lowers the temperature at which reactions are possible. The deposition temperature can be reduced to the extent of 500°C or more by using plasma than the furnace system to obtain the dielectric layer of different materials.

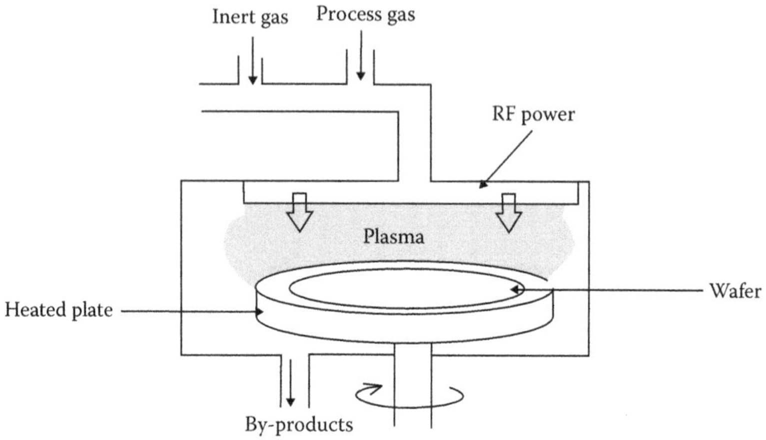

FIGURE 3.7
Plasma set-up of chemical vapour deposition.

The most universal technique to excite the plasma is the RF field. PECVD is mostly used to deposit insulating layer or dielectrics, and therefore, the DC excitation will not work in any way. Frequency range used here is usually from 100 kHz to 40 MHz. There are no constraints of using high vacuum, so the pressure is not reduced to a minimum value; it is set in the range of 50 mtorr to 5 torr. The ion density between 10^9 and 10^{11} $1/cm^3$ and average electron energies between 1 and 10 eV are generally maintained. Plasma CVD set is shown in Figure 3.7.

PECVD combines silicon with oxygen gas or nitrogen gas to create a plasma that deposits a thin film of silicon dioxide or silicon nitrate onto the substrate. PECVD is used in optics and microelectronics for the deposition of antireflective coatings, scratch-resistant transparent coatings, and deposition of passivation layers or dielectric layers, most importantly for deposition of etch-stop layers and also for encapsulation/protective coatings.

3.2.3 MOCVD

It is one of the sophisticated techniques for the deposition of epitaxial layer. The schematic is shown in Figure 3.8. Metal organic chemical vapour deposition (MOCVD) is widely used in compound/alloy semiconductor materials not applicable for elemental semiconductor. The example of alloy semiconductor is SiGe, which is mostly used in making heterostructure devices, e.g. high-speed transistor. Alloy semiconductor is also referred to as compound semiconductor. The most common example of compound semiconductor is AlGaAs. For single-crystal deposition of this material, MOCVD is used. It is basically a CVD technique used to form thin films. MOCVD uses a metal organic vapour which is used as a precursor (source gas) that reacts with

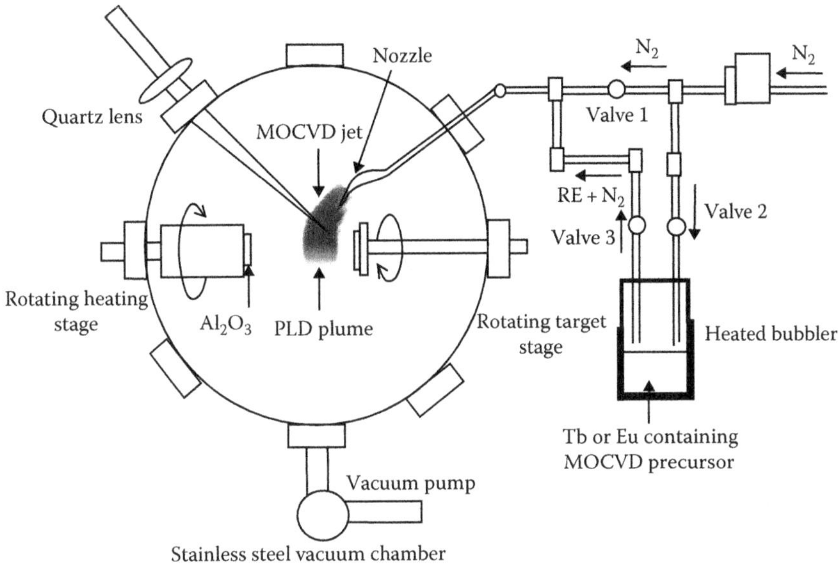

FIGURE 3.8
MOCVD set-up.

other reactants to form the desired film. The main advantage of MOCVD technique is that it is easy to remove volatile species from the reactor in order to avoid contamination for the next process step. It is always hard to remove solid contaminant or liquid contaminant. The excitation is generally thermal, but it may be plasma assisted.

This process differs from other processes in that it includes the transportation of the precursor molecules (III–V semiconductors) by carrier gas (N_2 or H_2 or both). Generally, N_2 with H_2 is preferred because it is difficult to produce pure nitrogen; it is always accompanied with O_2 molecule. These O_2 molecules can be eliminated by reacting with H_2 making water molecule. The substrate temperature is a crucial factor in this process; the substrate should be heated up prior to the deposition.

Group III–V semiconductors generally consist of the following:

Group III: metalorganic molecules (trimethyl or triethyl species), e.g. TMG, TMIn.

Group V: toxic hydrides such as AsH_3 and PH_3.

Volatile precursor molecules from high-pressure cylinder are mixed with carrier gas and transported by carrier gas. These metalorganic precursor molecules are generated from liquids of metalorganic, say TMG or TMIn, contained in a stainless steel bubbler kept in a constant temperature bath. High-purity liquid forms are the main sources of MO compounds. A fixed vapour pressure is maintained by keeping the temperature constant. H_2 flows through the bubbler and enters into the chamber where substrate is kept and the reactions take place. The flow of hydrogen is precisely controlled by mass flow controller. Each precursor molecule diffuses independently into the substrate until it comes into contact.

In case of GaAs deposition, organometallics used in MOCVD are TMGa (trimethyl gallium) and AsH_3 (arsine). Methyl group reacts with hydride yielding GaAs, H_2 and CH_4.

Reactions

$$Ga(CH_3)_3 + H_2 \rightarrow GaH(CH_3)_2 + CH_3$$

$$AsH_3 + CH_3 \rightarrow AsH_2 + CH_4$$

$$GaH(CH_3)_2 + AsH_2 \rightarrow GaAs + H_2 + CH_4$$

This is the basic principle of compound semiconductor deposition process in MOCVD technique. The major problem of MOCVD process is carbon contamination. Inside the chamber, the configuration is similar to CVD configuration. The primary factors of a VLSI process are low processing temperature, uniform step coverage, defect-free layer and batch production technique with reduced cost. All these requirements are mostly achieved by low-pressure deposition techniques like MOCVD.

3.3 Metallization

3.3.1 Different Types of Metallization

The last step in micro/nanodevice fabrication is to take an electrical output from what has been fabricated. Metal layers are deposited on the wafer to form conductive pathways. The most common metals that are used include aluminium, gold, silver, titanium, platinum, tungsten and copper.

Metallization is a very important step in VLSI technology. There are two types of metallization depending on the purpose of application. One is ohmic contact metallization in which the contact electrode makes an ohmic contact with the surface layer, and the second one is Schottky contact metallization.

The thumb rules for making two types of contacts are as follows:

- Metal on lightly doped silicon – rectifying Schottky contacts
- Metal on heavily doped silicon – low-resistance ohmic contacts

An ohmic contact is defined as one in which there is an unrestricted flow of majority of the carriers from one material to another. It is mandatory to heavily dope the Si regions N^+ or P^+ so that an ohmic contact is ensured. The silicon doping level should be high enough ($10^{19}/cm^3$) to make a good ohmic contact. If the doping level is high enough, the depletion region width is low enough so that charge carriers can tunnel through the barrier. Barrier thickness is related to depletion by the depletion region width in the semiconductor (which is proportional to 1/ND). Aluminium is generally used to make an ohmic contact between metal and silicon.

But in practice, ohmic contacts always show slightly non-linear (diode) I–V characteristics because metal and semiconductor work functions can never be exactly equal due to broken bands, impurities, etc. Jee et al. [2] have modified the surface of the ZnO thin films by a thin InSb layer by using a thermal evaporator to increase the work function without altering the physical properties of the film. InSb layer with a high work function could achieve the ohmic contact between ZnO and Pt. Ohmic contact has several positive characteristics as follows:

1. As the regions are heavily doped, the contact resistance in both N^+ and P^+ regions will be low.
2. Compatibility with Si processing technology (cleaning, deposition, etching, etc.).
3. Most importantly, there is no diffusion of the metal through the oxide layer.
4. No chemical reaction observed with the underneath layer.
5. Less effect on the electrical characteristics of the shallow junction.
6. Repeatability and stability in the long run.

In a lightly doped semiconductor, the situation is different: if the semiconductor doping level is low ($<10^{19}/cm^3$) enough, tunnelling is not possible, and the charge carriers will have to overcome the barrier height by thermionic emission (Figure 3.9).

$$\text{Barrier height is proportional to } (\phi_{metal} - \chi_{semiconductor})$$

where

ϕ_{metal} = metal work function

$\chi_{semiconductor}$ = semiconductor electron affinity

When the Fermi energy of the semiconductor is lower than the Fermi energy of the metal and they come in contact, under equilibrium, it is again the same on both sides. Electrons move from the metal to the semiconductor. These cause the bands to bend up near the metal–semiconductor interface. At the ohmic contact, the conduction band is lower than the Fermi energy near the metal–semiconductor interface. The semiconductor behaves like a metal in

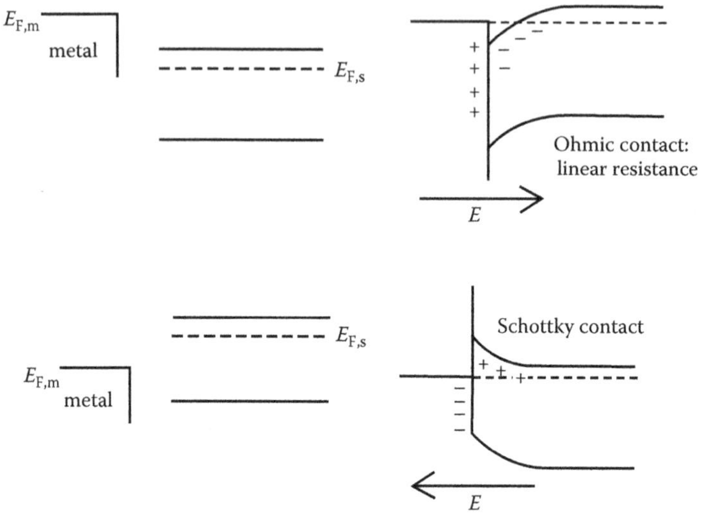

FIGURE 3.9
Ohmic contact and Schottky contact with n-doped semiconductor.

this region, and so an ohmic contact with a linear resistance occurs. In this case, the electric field helps to push the electrons towards the metal.

- Most of the catalytic noble metals (Pd, Pt, Rh, etc.) make Schottky junctions with the semiconducting metal oxides and provide the catalytic effect as well as the collection of carriers.

When the metal and the n-doped semiconductor come in contact, it gains the equilibrium by making the Fermi energy same on both sides. Electrons move from the semiconductor (lower work function) to the metal (higher work function). The Fermi energy level of the semiconductor moves downwards gradually until it is linear. That is the reason why the bands bend up near the metal semiconductor interface. A potential barrier exists across the junction, which is shown in the figure by immobile ions, and a depletion width occurs.

3.3.2 Methods

Metallization is generally accomplished with a vacuum deposition technique. Vacuum metallization is a form of PVD, a process of combining a metal with a non-metallic substrate through evaporation. The metallizing chamber is first evacuated to a predetermined vacuum level suitable for the evaporation of that particular metal. When the evaporated metal comes in contact with the target, the vapour condenses and creates a uniform layer of the evaporated metal.

The common deposition processes include thermal evaporation, e-beam evaporation, induction evaporation and sputtering. All are vacuum metallization process.

3.3.2.1 Filament Evaporation

Filament heating, also called thermal evaporation, is the simplest method of metallization. It is accomplished by the resistive heating of a filament which is kept inside a bell jar. As the temperature rises, the metal kept inside the filament melts gradually. As the current through the filament is increased, the heat generated by the filament vaporizes the entire metal until it goes off completely from the filament. The metal vapour condenses when it comes in contact with the substrate, forming the conformal metal layer.

3.3.2.2 Electron Beam Evaporation

Electron beam evaporation, frequently called 'e-beam', uses a focused beam of electrons. This beam of electrons heats the metal which is kept in a crucible and exposed for some time in that condition causing it to vaporize and condense on the wafers.

Time is one of the important parameters in this case because the thickness of the deposited film can be controlled in an effective manner.

3.3.2.3 Induction Evaporation

Induction evaporation uses RF radiation to melt the metal kept in a crucible. The metal is then evaporated and deposited on the wafers as discussed previously.

3.3.2.4 Sputtering

Sputtering is carried out in an inert gas atmosphere. Ions of an inert carrier gas are introduced into the chamber of low-vacuum atmosphere. The ions are produced by the application of an electric field to the atoms. Carrier gas is used to draw the ions in one place of the chamber where the target is fixed. The target is carrying the metal used for deposition. When the ions strike the target, metal atoms are displaced. The dislodged atoms are then deposited on the silicon substrate facing the target forming a thin film. Both dc and rf voltage can be used for sputtering.

3.3.3 Wire Bonding

Wire bonding is the final step of microelectronic fabrication and is used for electrical interconnection using thin (1 mil) wire (e.g. Au) and with the combined effect of heat, pressure and ultrasonic energy. It is analogous to welding in the subject of mechanical engineering and involves the joining of any two (or more) substances together to make the final device. The two surfaces are brought to intimate contact, and then inter-diffusion of atoms takes place. Heat is used to make this interatomic diffusion faster. So, the bonding time is reduced. Wire bonding is also used to connect different components on the same substrate, such as resistor and capacitor.

Ultrasonic energy is used to improve the surface smoothness. As pressure is mandatory for wire bonding, naturally this would lead to the deformation of the contact pad material. This deformation is partially covered up by the inclusion of ultrasonic energy. Pressure is used to make the contact between the bonding material and the contact pad.

There are three types of wire bonding processes depending on the types of energy used. These are as follows:

1. Thermocompression
2. Thermosonic
3. Ultrasonic

There are basically two forms of wire bond: ball bonding and wedge bonding. These two have been shown in Figure 3.10.

Ball bonding is again of two types, thermocompression (T/C) and thermosonic (T/S). T/C utilizes pressure and temperature to form a bond, and in T/S, bonding is the amalgamated effect of thermocompression (T/C) and thermosonic (T/S) and also ultrasonic energy in the process. Ball bonding is generally performed with pure gold wire (typically for 1 mil gold wire) drawn out from a gold wire bobbin as it can be easily deformed with applied pressure. In both methods, the end of the bond wire is transformed into a ball shape by the application of an electronic flame off. The ball is then positioned by a mechanical shaft, just above the bond pad which lies on the substrate, and contact is made by any of the earlier two methods. Aluminium is

(a) (b)

FIGURE 3.10
SEM image of (a) gold ball bond and (b) aluminium wedge bond. (Courtesy of West Bond Inc. website. [Online]. Available: http://www.westbond.com/.)

much more susceptible to oxidation, that is why it is not used in ball bonding. The presence of heat at the time of ball formation can create oxides that will positively obstruct the joining process.

Ball–wedge bonding is the most common method of wire bonding in which the first bond, i.e. from where it is started, is the ball bonding, and the second bond, i.e. at where it is terminated, takes the shape of a wedge.

The ultrasonic wedge bonding process is generally performed with aluminium wire. The welding bonding process deforms the wire into a flat elongated shape of a wedge; the bond pad is made of either aluminium or gold. Pure aluminium wire is generally avoided as Young's modulus is too low to form into a fine wire; therefore, it is often alloyed with small quantity of silicon or magnesium to make its Young's modulus high. Gold and copper can also be used as a wedge bonding material because of its high strength and stiffness. The main advantage of these two materials is that it can be bonded even at room temperature.

Another term existing in the bonding process is wedge–wedge bonding; in this case, the first bond does not contain a ball; it is wedge shaped, and the second ball is also wedge shaped, which is why this wire bonding procedure is called wedge–wedge bonding.

References

1. C. Henchy, U. Kilmartin and G. McCaffrey, Vacuum ultraviolet spectroscopy and photochemistry of zinc dihydride and related molecules in low-temperature matrices, *J. Phys. Chem. A* 117, 9168–9178 (2013).
2. S. H. Jee, N. Kakati, S. H. Lee, H. H. Yoon and Y. S. Yoon, Ohmic contact between ZnO and Pt by InSb layer in a ZnO Schottky diode, *Appl. Phys. Lett.* 98, 142108 (2011).

4

Photolithography: Pattern Transfer

4.1 Introduction

Photolithography is the photographic technique to transfer replica of a master pattern into a substrate of a different material (usually a silicon wafer). In the case of silicon, a SiO_2 insulating layer is used to cover the substrate. A thin layer of an organic polymer, which is sensitive to ultraviolet radiation, is then deposited on the oxide layer; this is called a photoresist (PR). A photomask, which consists of a transparent plastic coated with a chromium pattern (opaque), is placed in contact with the PR-coated surface. The masks are of two types: dark film mask and bright film mask. The wafer is exposed to the ultraviolet radiation transferring the pattern on the PR, which is then developed by an organic developer. The ultraviolet (UV) radiation causes a change in the polymeric bond in the exposed areas of the PR. PRs are of two types: positive and negative. Positive PRs are strengthened by UV radiation, whereas negative PRs are weakened. While developing, the developer removes either the exposed areas or the unexposed areas of PR leaving a pattern on the oxide-coated wafer surface. The resulting PR pattern is either the positive or the negative image of the original pattern of the photomask.

A chemical (usually hydrochloric acid) is used to remove the extra metal other than the pattern outside the exposed areas of the PR. The remaining PR pattern is subsequently removed, usually with acetone for positive PR and nitric acid for negative PR pattern without damaging the oxide layer on the silicon, leaving a metallic pattern on the silicon dioxide surface.

Inorganic PR is presently capturing the market due to nanoparticle (254 and 193 nm) patterning.

In the conventional 193 nm, PR nanoparticle core is attached with functional groups. Many resist materials degrade in plasma or ion beam environments.

The prime limitation in optical lithography is diffraction. Typically, electron beam lithography does not suffer from diffraction limitation. In the case of optical lithography, it can also be minimized by the use of excimer lasers that generate nanosecond pulses and very fine patterns, modifying the PR chemistry and by introducing resolution enhancement techniques.

4.2 Photoresist for Structuring

Before applying PR, surface preparation is a vital factor to promote the resist adhesion factor. In view of that, after cleaning, adhesion promoters are used to assist resist coating.

Resist adhesion is affected by the following factors:

- Moisture content on surface
- Wetting characteristics of resist
- Type of primer
- Delay in exposure and prebake
- Resist chemistry
- Surface smoothness
- Stress from coating process
- Surface contamination

Ideally, the wafer surface should be completely free from water molecule or moisture. It is very much essential to keep the wafers in a heating oven prior to priming and coating. It should be kept for 15 min at 80°C–90°C in the convection oven (muffle furnace).

4.3 Some Important Properties of Photoresist

4.3.1 Sensitivity

Being an organic polymer, PR is very much sensitive to radiation and alters its chemical properties drastically. It is even sensitive in normal light of operation. That's why it is always kept in the dark room. The high sensitivity in turn reduces the exposure time, which ultimately greatly affects the fabrication cost. For optical lithography, the resists (either positive or negative) respond to 330–430 nm.

Recently, deep sub-micrometre lithography is of great interest as it utilizes the deep UV region (150–300 nm) as these shorter wavelengths are capable of producing higher resolution that can be helpful for the miniaturization of the device. But there are some constraints regarding the selection of wavelength as the operation in these shorter wavelengths needs a vacuum environment; hence, the region of practical interest is ≥200 nm, which does not require the vacuum strictly. This range is again inhibited by the availability of high-intensity deep UV sources. The most practical source today is an Hg–Xe lamp that has a peak output of 250–290 nm. Excimer lasers are used as sources for the deep UV range.

The PR also affects the functionalization of e-beam lithography as well as x-ray lithography (5–50 Å). For the e-beam technique, the resist should be sensitive to electron irradiation in the 10–30 kV range for reticle mask generation.

4.3.2 Adhesion

The adhesion capacity of resist should be high enough to the underlying substrate so that it does not detonate during subsequent processing. This is a matter of concern for wet etching only as in the case of dry etching, the fabrication process involves plasmas or etchant gases to remove the substrate material.

4.3.3 Etch Resistance

High etch resistance is an essential parameter for effective pattern transfer. Etch resistance of a PR depends on the ratio of the unsaturated atomic components of the PR polymer structures; this unsaturation offers the optimal level of etch resistance. This unsatisfied bond originated from the double bonds or cyclic structures present in the PR polymer. The aromatic units present in 248 nm PR polymers present a high degree of unsaturation due to the three double bonds and one ring that comprises each phenyl group. These characteristics are suitable to provide resistance to plasma etching conditions. The etch rates of the resists strongly depend on the etching conditions. The etching equipment has an influence, the amount of open wafer surface to be etched, the etch gas composition and all other parameters such as pressure, temperature and voltage. PRs withstand strong acids very well. No erosion of the resist film is observed.

Concentrated hydrofluoric acid (HF) is challenging for all PRs. Strong oxidizing acids can also cause some problems. The resist stability depends on the temperature and the composition of the etchant in such cases. HF etching is a bit demanding. HF does not attack the resist, but it can diffuse under the PR and lift it from below causing bad adhesion of the resist on the substrate. This is why a film thickness as high as possible should be chosen, and the resist should be hardened (stronger prebake + hard bake). Nevertheless, it depends strongly on the HF concentration and the etch time, how well the PR sustains the etching.

4.3.4 Bubble Formation

There are several reasons for the formation of bubbles in the resist. Big bubbles could arise if the resist was shaken too much or if the resist has undergone a bigger temperature change. This applies mainly for resists with higher viscosity. Bubbles occurring during the prebake of thick resist films are caused by solvent evaporation. It is important to maintain a certain relaxation time between spin coating and prebake to avoid this. Only high-viscosity resists show this tendency. Factors contributing to bubble formation are as follows:

- Insufficiently softbaked (=prebaked)
- High exposure dose
- Use of too much primer (hexamethyldisilazane [HMDS] on Si and SiO_2)
- Especially a too high humidity

4.4 Types of Photoresists: Negative and Positive Photoresists

4.4.1 Negative Photoresists

In negative PRs, the monomers undergo cross-linking and polymerize upon exposure to light, and the exposed parts become insoluble in the developer, usually in an organic solvent. The resist becomes hardened upon exposure to light. It could not be dissolved and washed away in solvents. This type of resist is known as negative resist because the replica that has been formed in it is the inverse of the object.

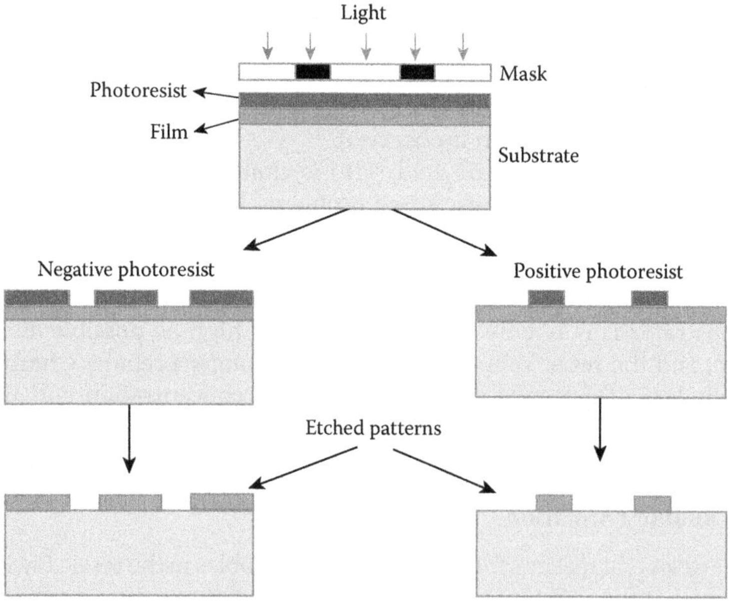

FIGURE 4.1
Patterning with photoresist.

4.4.2 Positive Photoresists

The development of positive resist is comparatively recent. The replica that has been formed is an exact image of the object. The material is initially insoluble, but degrades upon exposure to radiation. Generally, in the case of positive resists, an acidic group is generated on photolysis, and the basic developer easily washes this out (Figure 4.1).

Both positive and negative resists are used in microfabrication process depending on the requirement. The most commonly used positive resist consists of the photoactive compound (PAC) and a matrix material called resin. Upon exposure to UV light, the PAC undergoes one or more chemical changes. The exposed-resist areas turn either more soluble or less soluble depending on the type of resist and mask used than the unexposed-resist areas. With the use of appropriate developer, which is basic in nature (KOH, NAOH, TMAH, etc.), this change in solubility induced by the photochemical change allow the formation of the image.

4.5 Designing of Mask Layout

A mask for optical lithography consists of a transparent plate called blank, covered with a patterned film of opaque material. The blank is made of soda lime, borosilicate glass or fused quartz. The advantage of the quartz is that it is transparent to deep UV (≤365 nm) and has a very low thermal expansion coefficient.

In a latest optical-mask-making process, a computer tape is used to drive an optical pen that directly writes the pattern on the reticle mask. Computer-generated patterns are used as very large scale integration (VLSI) masks. Then, the layout is converted into a reticle or master mask. Several working masks have been generated from this master mask by contact printing technology though the master mask can be used directly for pattern transfer onto the substrate. For pattern generation, the most popular one is e-beam writing technique due to its lower diffraction complexity. Owing to the diffraction limitations, optical techniques are still in the market due to its simpler design tool.

Very recently, a new technique has been developed in which mask-making step can be avoided by a simultaneous procedure, i.e. pattern generation and patter transfer can be done in a single step, which means writing pattern directly on the wafer through the instrument used for this is highly expensive and the overall fabrication process would get costlier.

Computer-aided design (CAD) is generally used in MEMS for the design of photolithographic masks. This is a straightforward process as MEMS structures are relatively large in comparison to the sub-micrometre structures usually associated with silicon chip components. In addition to using CAD for mask design, CAD and finite element analysis are important simulation tools for the design of MEMS applications. Unfortunately, to date, there is a lack of adequate advanced software-based design tools to fully model, analyse and simulate MEMS microstructures as well as integrated MEMS/IC devices. This has acted as a barrier to the development of MEMS devices and systems.

One of the most successful and commercially available software design tools today is MEMCAD, a package from Microcosm Technologies in North Carolina, United States. The MEMCAD system defines device layout and process, constructs the 3D geometry of the device, assembles a detailed 3D model and analyses device performance as well as device sensitivity to manufacturing and design variations.

MEMS Pro, a package from Tanner Research in California, enables designers of MEMS to simulate the growth/deposition, implantation/diffusion and etch steps in an MEMS fabrication process.

4.6 Photolithography Process

The 10 basic steps of photolithography are as follows:

1. Surface preparation
2. PR application
3. Prebake (soft bake)
4. Align and expose
5. Develop
6. Postbake (hard bake)
7. Inspection
8. Etch
9. Resist strip
10. Final inspection

4.7 Application of Photoresist and Prebake

Photoresists are nothing but the photo-reactive materials consists of carbon polymer. Wafer is held on a spinner chuck by vacuum, and resist is coated to uniform thickness by spin coating. Typically, 3000–6000 rpm for 15–30 s is generally used in microdevice fabrication.

Resist thickness is set by the following:

- Primarily, resist viscosity
- Secondarily, spinner rotational speed

Resist thickness is given by $t = kp^2/w^{1/2}$, where

- k is the spinner constant, typically 80–100
- p is the resist solids content in percent
- w is the spinner rotational speed in rpm/1000

Mostly, resist thicknesses are 1–2 mm for commercial Si processes. Prebake is used to evaporate the coating agent and solidify the resist after spin coating. Typical thermal cycles used for this densification are as follows:

- 90°C–100°C for 20 min in a convection oven where solvent at the surface of resist is evaporated first, which causes the resist to develop impermeable skin, trapping the remaining solvent inside. Heating must go slow to avoid solvent burst effects.
- 75°C–85°C for 45 min on a hotplate. Hotplating is generally preferred as it does not trap solvent as it does in the oven, but the dust particle in the air is another issue for hotplating. Moreover, temperature rise starts at the bottom of the wafer and works upward, more thoroughly evaporating the coating solvent. It is much faster and more suitable for automation.

Microwave heating or IR lamp heating is also used for commercial purposes for large-scale production. The thickness of resist is decreased to 25% after prebake for both resists.

4.8 Alignment, Exposure, and Pattern Formation

UV light is used because as per the modern microelectronics technology, the device size is so small that the wavelength of the exposed light is a limiting factor. The wavelengths that are generally used are 436 nm (G line), 405 nm (H line), 365 nm (I line) and 248 nm (called deep UV); at present, even shorter-wavelength UV and also x-ray are still on the way, but it suffers from workers' health risk. G line and I line resists generally contain three components: inactive resin, PAC and solvent. The lithographic technique is shown in Figure 4.2.

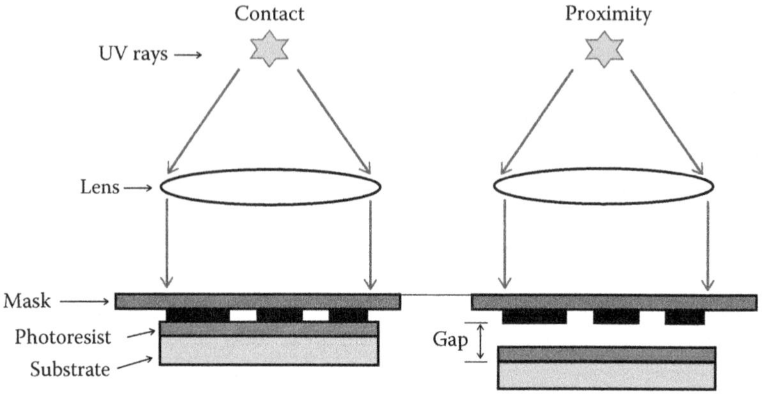

FIGURE 4.2
Optical lithographic technique.

Resist thickness may vary across the wafer, which may lead to under- or overexposure resulting in variation in line width.

4.9 PR Developer and Postbake

This process is used to stabilize and harden the developed PR prior to the processing step that will mask the resist. The main parameter in this process is the plastic flow or glass transition temperature. Postbake removes any remaining traces of the coating solvent or developer. This eliminates the solvent burst effects in vacuum processing and introduces some stress into the PR. Some shrinkage of the PR may occur in this process. Longer or hotter postbake makes resist removal much more difficult.

4.10 Stripping (Photoresist Removal)

This process is the removal of the PR and any of its residues. Simple solvents are generally sufficient for non-postbake.

Types of PRs are as follows:

- Positive PRs
    - Acetone
    - Trichloroethylene (TCE)
    - Phenol-based strippers (Indus-Ri-Chem J-100)

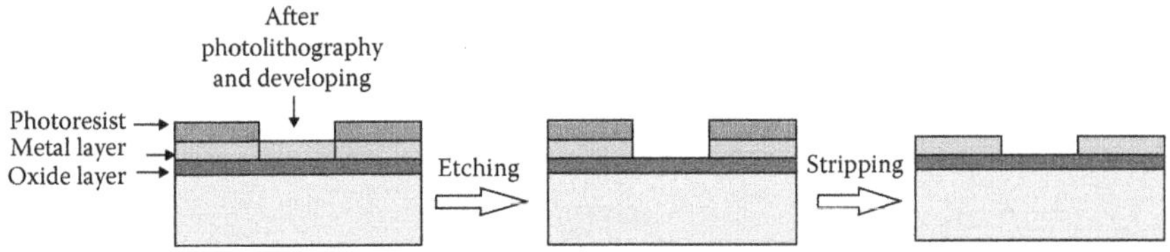

FIGURE 4.3
Stripping method.

- Negative PRs
  - Methyl ethyl ketone (MEK), $CH_3COC_2H_5$
  - Methyl isobutyl ketone (MIBK), $CH_3COC_4H_9$

Plasma etching with O_2 (ashing) is also effective for removing organic polymer debris. Also Shipley 1165 stripper (contains n-methyl-2-pyrrolidone) is effective on hard postbaked resist. This process is the removal of the PR and any of its residues (Figure 4.3).

4.11 Some Advanced Lithographic Techniques

4.11.1 Electron Beam Lithography

The set-up is shown in Figure 4.4. This electron beam lithography is very much similar to the scanning electron microscope technique with the addition of beam blanking and computer-controlled deflection facility. There are several more extra features: one is laser-driven interferometer and another is fiducial mark detector. This mark detector facilitates the exact positioning of the system for each successive pattern transfer operation. The area of the deflection field is typically 2 mm × 2 mm. To cover the mask altogether, a number of such field should be stitched together to form the entire pattern.

Two types of scan systems are in use: the raster scan and the vector scan. In the raster scan system, the scan is done by covering the rectangular strips one by one and dividing the entire circuit in order to form the complete chip pattern. In a vector scan system, the e-beam is controlled to scan a feature and almost immediately moves to the next feature and continues.

Electron beam lithography has several advantages:

- Direct patterning without a mask
- Production of micrometre and sub-micrometre resist geometries
- Precise control of operation
- Better depth of focus

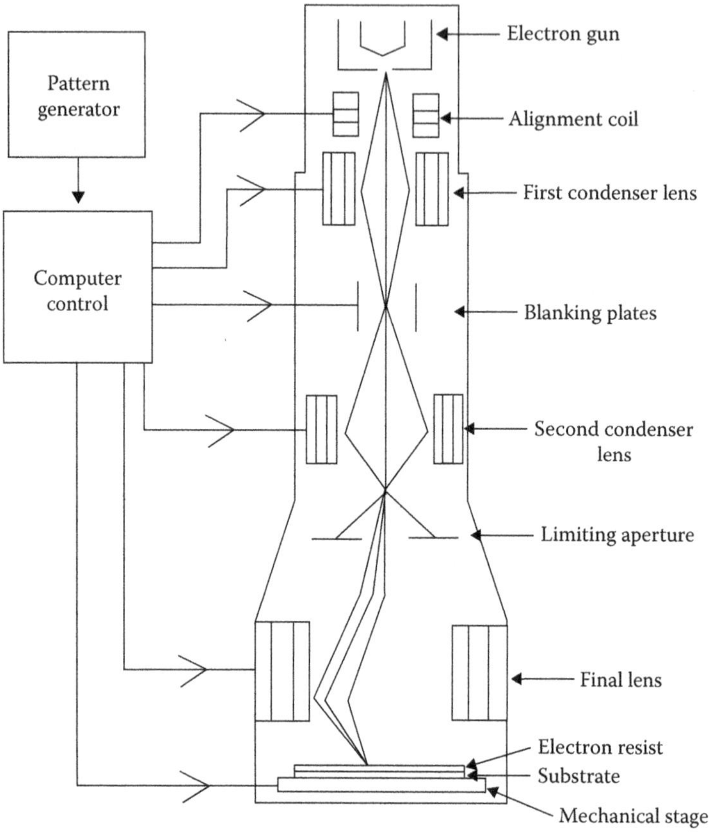

FIGURE 4.4
Electron beam lithography set-up.

There are also several disadvantages. It gives low throughput (approximately 5 wafers/h at less than 0.1 μ resolution). This process is highly automated; therefore, any type of failure will hamper the production. Moreover, electron scattering is a problem in this lithographic technique as the resolution depends on electron scattering and the size of the beam. Diffraction has negligible effect on resolution, which should be noted. Electron lithography is best suited for the production of photomasks.

4.11.2 Ion Beam Lithography

Scattering of electrons has severe effect on resolution. Therefore, it was thought that the use of ions instead of electrons in the lithographic process can minimize the issue. To overcome this limitation, ion beam lithography is introduced in the market (Figure 4.5).

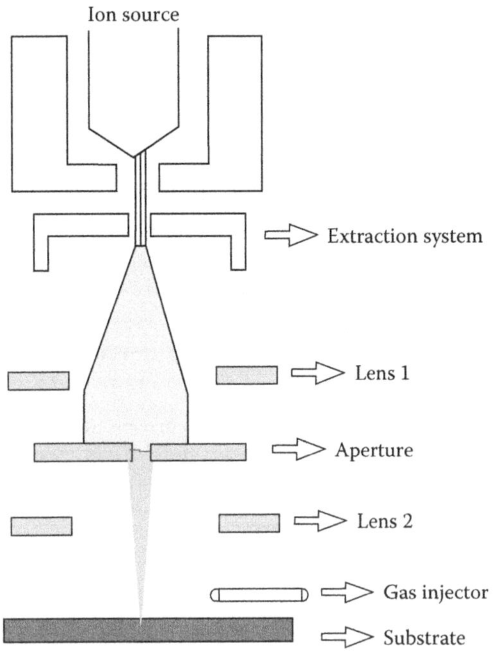

FIGURE 4.5
Ion beam lithography.

While penetrating the resist, ion beam does not suffer from scattering as electron beams. In addition, secondary electrons produced by ion bombardment are very weak and have a least effect on the scattering phenomenon. The minimum feature size is set by these scattering processes. Next, the ion sensitivity of resists is a function of the ion energy and mass; it is much greater than that of electrons. Thus, ion beam techniques are practically more rapid than e-beam techniques as the structural property of resist changes fast when exposed to ion beam. Ion sources that are generally used are radio frequency ion sources, i.e. H^+, He^+ and Ar^+ in the 100 keV range with focusing optics to image a field of 5 mm × 5 mm in a single exposure. Direct writing using this technique is not done yet; it is under investigation.

4.11.3 X-Ray Lithography

The diffraction problem can be totally eliminated if a shorter wavelength is used. X-ray lithography is one of the major solutions to the earlier problem. This approach utilizes the benefit of extremely short wavelengths, which is possible with x-ray sources. Fabrication of sub-micrometre-level structures at lower cost is made possible by this technique which is competitive with

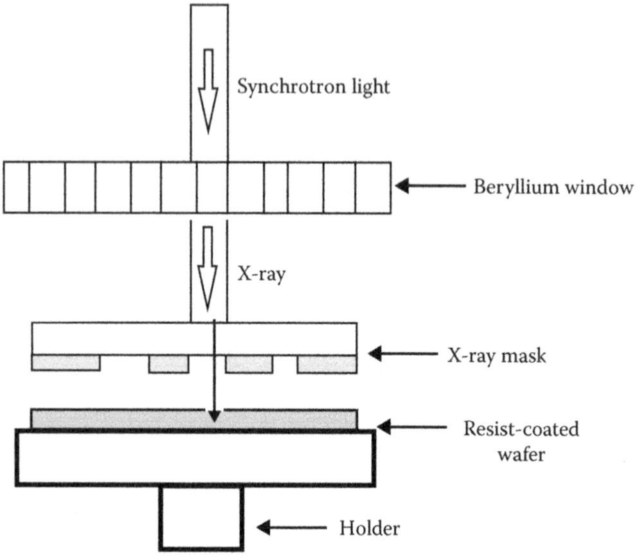

FIGURE 4.6
X-ray lithography.

optical and UV lithographic techniques. The physics behind it is totally different from the conventional lithographic technique. The optical technique uses parallel light and focuses it, whereas x-ray systems are restricted to a point source (Figure 4.6).

Mask is the most critical element in x-ray lithography system due to the unavailability of the PR as high-energy x-ray destroys the resist property.

4.11.4 Phase-Shift Lithography

Diffraction is the most encountered problem in most of the photolithographic techniques. Mostly, higher frequencies get corrupted due to diffraction for which sharp features are lost. This problem is much pronounced when there are more than one mask pattern. It becomes undistinguishable.

This difficulty can be avoided by the use of phase-shift photolithography. It is basically a resolution enhancement technique using an optical phase-shifting method. This is based on the already established formula that the resolution of an optical system can be improved by increasing the numerical aperture and reducing the wavelength though there is a compromise with the depth of focus. Variation of the wavelength does not justify the PR composition, so a totally new optical lithographic system is required.

Phase-shifting technology enables sufficient image contrast to successfully print significantly smaller silicon features by optical phase-shifting method and is shown in Figure 4.7.

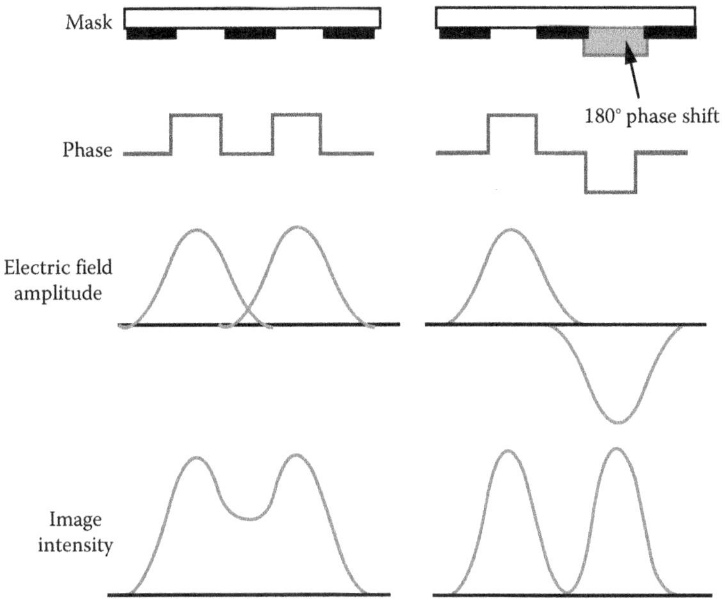

FIGURE 4.7
Phase-shift lithography.

The figure shows two adjacent slits in an optical mask. The light intensity from each of these slits is diffracted into the neighbouring regions that surround them. If these slits are made closer, in normal optical lithography, the diffracted regions interfere with each other until the individual intensities are resolved. There are a number of disadvantages associated with optical lithographic technique; still it is widely used, and it is a promising method in the microfabrication process as it has high throughput, has good resolution, is economical and is easy to operate.

5

Structuring MEMS: Micromachining

5.1 Introduction

MEMS stands for microelectromechanical system. Toxic and hazardous gas detection using gas sensors based on MEMS technology [1–4] is a well-known phenomenon in today's environment. An integrated gas sensor with inbuilt microheater is necessary in most gas sensors because the chemical reaction involved in the sensing layer with the target gases takes places at some elevated temperature. A microheater array is very promising in enhancing the selectivity of the gas sensor with respect to sensor selectivity, with a compromise that the use of the sensor array also leads to an increased size of the device. If separate gas sensors are used to detect individual gases, the excessive power consumption will lead to rejection of the device. In the past years, the earlier-mentioned difficulties led to innovative substrate technology and extensive investigation for suitable sensing materials and signal evaluation. Power consumption to increase the operating time of a device is a major issue for a battery-operated instrument. The integration of thin gas-sensitive metal oxide layers in standard microelectronic processing can be done with the use of micromachining steps to fabricate micromachined metal oxide gas sensors. The advantages of MEMS technology over conventional ceramic sensors are mentioned in the following:

1. The sensing layer is deposited onto a thin dielectric membrane of low thermal conductivity, which provides good thermal isolation between the substrate and the active area of the membrane. In this way, thermal loss of the device can be minimized and the power consumption can be kept low (typical values obtained lie in the range between 30 and 150 mW) [5,6].
2. The surrounding silicon in the MEMS-based gas sensor behaves as a support of the device, and as a result, mounting of the sensor element becomes much easier than for an overall hot ceramic sensor element. The use of silicon facilitates that the signal processing electronics can be integrated on the same substrate.

3. Extreme miniaturization is possible. A minimal spacing between the fingers of interdigitated electrodes lying in the nm range is possible [7]. The use of IDE enhances the sensitivity of the gas sensor. For this reason, the sensing area or the active region can be tremendously reduced, and moreover, the use of interdigitated electrodes with a high length-to-width ratio allows the evaluation of sensing films with very high sheet resistance.
4. The small thermal mass of each micromachined membrane in a microheater array allows rapid thermal programming, i.e. the time required to attain that predetermined value of the temperature due to the fast surface process kinetics. It was found that the power consumption of the device is proportional to the size of the heated area. As a matter of fact, a smaller heated area means less power consumption and faster response time.

The terms 'MEMS' and 'micromachining' are inseparable. To make an MEMS device, micromachining is an essential step. Micromachining is a process to fabricate the micromechanical parts using silicon. There are several positive aspects of silicon micromachining. It is inexpensive as it can be batch fabricated, the performance of the device is higher, and it can easily be interfaced with the signal conditioning unit.

Micromachining of the silicon substrate is done by selective etching of the substrate to leave behind the desired geometries for the micromechanical structures. There are several different types of micromachining technologies, which are discussed next.

5.2 Bulk Micromachining

Bulk micromachining is a fabrication technique that builds mechanical elements by starting with a silicon wafer, then etching away unwanted parts and being left with useful mechanical devices. Typically, the wafer is photo patterned, leaving a protective layer on the parts of the wafer that you want to keep. The wafer is then submersed into a liquid etchant, like potassium hydroxide, which eats away any exposed silicon. This is a relatively simple and inexpensive fabrication technology and is well suited for applications that do not require much complexity and that are price sensitive.

Today, almost all pressure sensors are built with bulk micromachining. Bulk micromachined pressure sensors offer several advantages over traditional pressure sensors. They cost less, they are highly reliable and manufacturable, and there is very good repeatability between devices. All new cars on the market today have several micromachined pressure sensors [2], typically used to measure manifold pressure in the engine. The small size and high reliability of micromachined pressure sensors make them ideal for a variety of medical applications as well.

Bulk micromachining can be done in two ways.

5.2.1 Wet Etching

Wet etching describes the removal of material through the immersion of a material (typically a silicon wafer) in a liquid bath of a chemical etchant. Materials are dissolved in the chemical solution.

In dry etching, materials are removed from the bulk by vapour phase etchant in normal cases.

The mechanisms for wet chemical etching involve three essential steps (Figure 5.1) in the following:

1. The reactants are transported by diffusion to the reacting surface.
2. Chemical reactions occur at the surface.
3. The products from the surface are removed by diffusion.

The etch rate is nothing but the amount of film removed per unit time and is highly influenced by agitation and temperature. Etch can also be done by spraying etchants into the wafer surface or immersing the wafer into the chemical etchants as shown in Figure 5.2.

Etching should be uniform across a wafer. Etch rate density uniformity is given by the following equation:

$$\text{Etch rate uniformity (\%)} = (\text{maximum etch rate} - \text{minimum etch rate}) / (\text{maximum etch rate} + \text{minimum etch rate}) \times 100\%$$

The main constraints of wet etching are that it requires a mask patterned accurately as the required structure. At the same time, it should also be checked to ensure that the mask material (e.g. SiO_2, Si_3N_4) will not get dissolved in the chemical solution.

Isotropic etchants etch the material in all directions simultaneously, the effect of which is the removal of the material under the etch mask also at the same etch rate as they etch the whole material; this undesired removal of the silicon is known as undercutting.

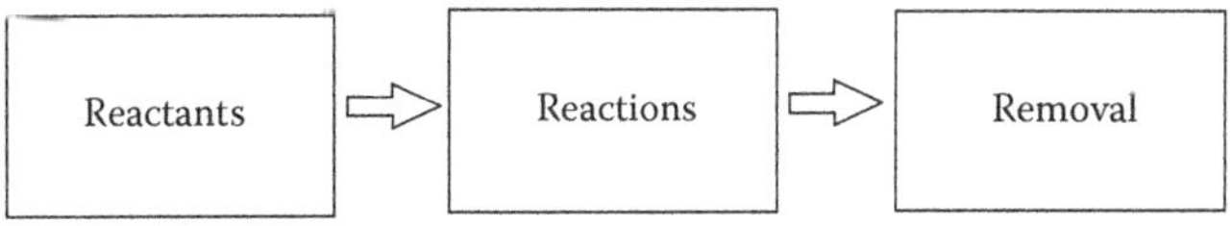

FIGURE 5.1
Basic steps of wet etching.

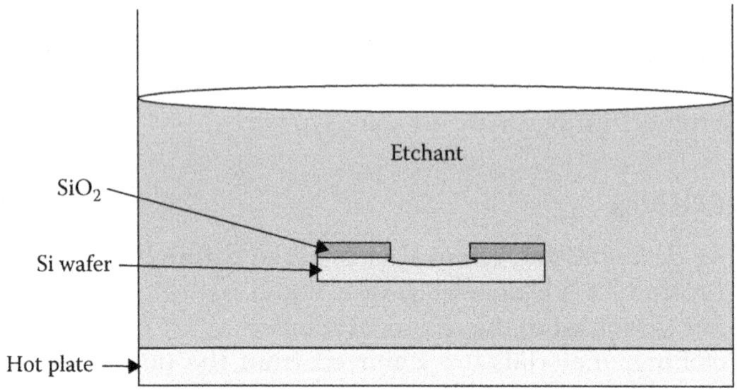

FIGURE 5.2
Wet etch set-up.

The most common isotropic etchant is HNA, which is a mixture of hydrofluoric acid (HF), nitric acid (HNO_3) and acetic acid (CH_3COOH). The limitation of the isotropic etchant is that they are limited by the geometry of the structure to be etched. Etch rates can be controlled in a desired manner and can also be stopped by diffusing an etch-stop layer of specific thickness.

Anisotropic etchants etch faster in a preferred direction. Potassium hydroxide (KOH) is the most common anisotropic etchant as it is relatively safe to use. Structures formed in the substrate are dependent on the crystal orientation of the substrate or wafer. Most such anisotropic etchants progress rapidly in the crystal direction perpendicular to the (110) plane and less rapidly in the direction perpendicular to the (100) plane. The direction perpendicular to the (111) plane etches very slowly if at all. Silicon wafers, originally cut from a large ingot of silicon grown from single seed silicon, are cut according to the crystallographic plane. They can be supplied in terms of the orientation of the surface plane. Dopant levels within the substrate can affect the etch rate by KOH and, if levels are high enough, can effectively stop it. Boron is one such dopant and is implanted into the silicon by a diffusion process. This can be used to selectively etch regions in the silicon leaving doped areas unaffected.

5.2.1.1 Isotropic and Anisotropic: Empirical Observations

Depending on the chemicals used and materials to be etched, this etching can be divided into two groups: isotropic and anisotropic.

Isotropic etching is generally avoided in bulk micromachining as it is not direction dependent. Structures with rounded side walls are produced using this etching. This has been used to make silicon waveguides [8]; sometimes anisotropic etching with isotropic pre-etch step is performed.

Wet etching can be isotropic, unless the material is crystalline, and the chemical etches in some directions more rapidly than in others. The common isotropic etchants are HF, HCl, etc., and are generally acid based.

The anisotropic etching of silicon is a key process in bulk micromachining as it is well suited to batch fabrication. Anisotropic etching means different etch rates in different directions in the material. One of the most common examples of this is the etching of an MESA structure in which anisotropic etchants of silicon etch the <100> and <110> crystal planes significantly faster than the <111> crystal planes. The etch rate for <110> surfaces lies between those for <100> and <111> surfaces. Another common example is that V groove structures, useful for the fabrication of silicon membrane, are easily fabricated using an anisotropic etchant like EDP, KOH, hydrazine and tetramethylammonium hydroxide (TMAH) [9–11].

Mostly used technology for bulk structuring for microsensors as

1. Thickness of the membrane can be well controlled.
2. Anisotropic etching was well suited to batch fabrication.

Specifications for the etched structures (such as high etch rate ratios of <110> and <100> to <111> planes, short etch times and minimum roughness) can be optimized by varying the etch parameters. For sensor applications, generally <100> silicon is used.

Figure 5.3a depicts the resulting structure of bulk micromachining by anisotropic wet etching of <100> silicon wafer. Etching continues along the <100> planes, while it is stopped along <111> planes. From the figure, it is evident that <111> planes make a 54.75° angle with the <100> planes. For rectangular or square window, the sides are aligned with the <110> direction, and hence, no undercutting has been observed. Figure 5.3b shows the isotropic etching of silicon. No such orientation has been found, that is why the term 'iso' appears.

The width of the bottom surface W_b is given by [8]

$$W_b = W_o - 2t \cot(54.75°) \tag{5.1}$$

where

W_o is the width of the etch mask window on the wafer back surface
t is the etched depth

If <110> oriented silicon is etched in KOH water etchant, essentially straight-walled grooves with sides of <111> planes can be formed. This is the exceptionality of this process that some planes grow while others disappear.

While selecting an etchant, etch rate, etch selectivity, degree of anisotropy and low toxicity should be encountered first. Etchants can be characterized by using the following characteristics:

Direction dependency (isotropic or anisotropic)

Etch rate

Anisotropic etch rate ratio (only for anisotropic etchants, 1:1 to 400:1 for <100>/<111> planes)

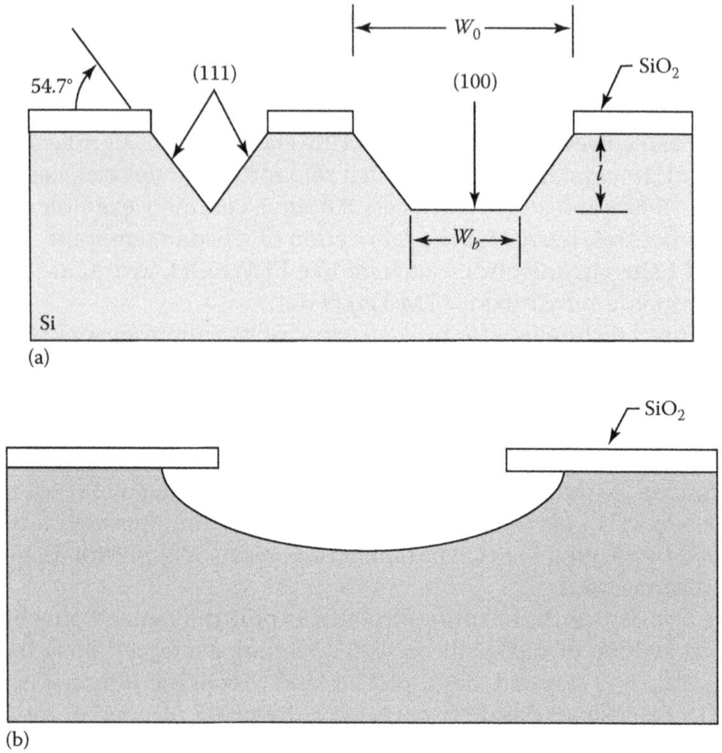

FIGURE 5.3
(a) Anisotropic and (b) isotropic etching of silicon.

Dopant dependence/selectivity

Temperature of etching (20°C–100°C)

- *Direction dependency*: The most important aspect in classifying one etchant is its ability to etch different crystal planes with different etch rates in a crystal lattice that is exposed to it. Isotropic etchants are direction independent resulting in rounded edges and corners, whereas anisotropic etchants are direction dependent.
- *Etch rate*: Etch rate is the function of temperature, chemical composition, volume and concentration of chemicals.
- *Anisotropic etch rate ratio*: Anisotropic etch rate ratio is the ratio of etch rates regarding different crystal planes and can vary in a wide range. This ratio can vary from 1:1 to 400:1 depending on the choice of etchant.
- *Selectivity*: Selectivity is one of the important attributes of an etchant. This is mainly related to the dopant dependency of

etchants. Some etchants can etch a particular material very selectively that they are exposed to; in that case, to control the etch depth, a layer of different materials can be used to stop the etching process or lower the etch rate by introducing a separate layer in the bulk. Somewhere, if this is not preferred, it is better to choose a non-selective etchant [10].

- *Etching temperature*: Etching temperature is a very important parameter as temperature in any mechanical body induces thermal stress. Thermal stress may generate microcracks or some defects in the system. In general, lower temperatures are preferred than higher temperature, as temperature-induced stress can be minimized.

5.2.1.2 Convex and Concave Corner Compensations

Wet etching is the best-suited process for etching the bulk substrate. The problem with substrate etching is that isotropic processes will cause undercutting of the mask layer by the same distance as the etch depth. On the other hand, as anisotropic etching is direction dependent, therefore, it allows the etching to stop on certain crystal planes and to continue on other planes, which in turn results in wastage of space produced (Figure 5.4).

With dry etching, it is possible to etch almost vertically without undercutting, which offers much higher resolution. The basic structure of single-crystal silicon has been shown in the figure. The atoms are oriented such that the {111} planes represent the highest packing density.

Due to its position, the surface atoms always have the unsatisfied covalent bond, which is known as the dangling bond. The dangling bonds easily react with the etching agent due to its unoccupied valence state. Theoretically, this is true, but practically, it does not remain idle; rather, the bonds are terminated with hydrogen atom in water. It is apparent that due to the high packing density of (111) plane, the number of dangling bonds is less in that plane. This is the reason of established stability of this plane against etching.

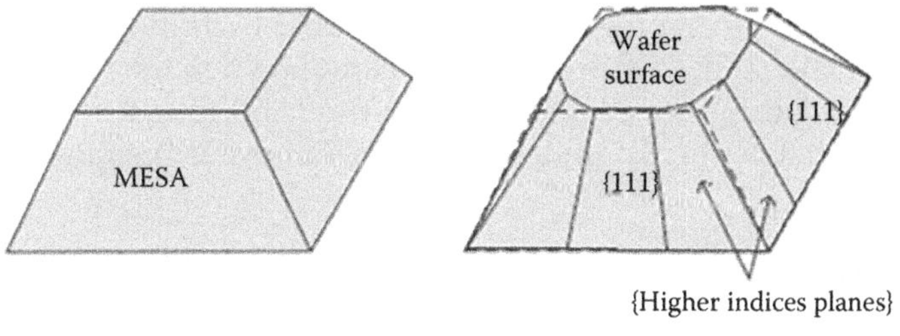

FIGURE 5.4
Convex corner undercutting of <100> silicon with MESA structure. (From Schroder, H. et al., *J. Microelectromechanical Syst.* 10, 88, 2001.)

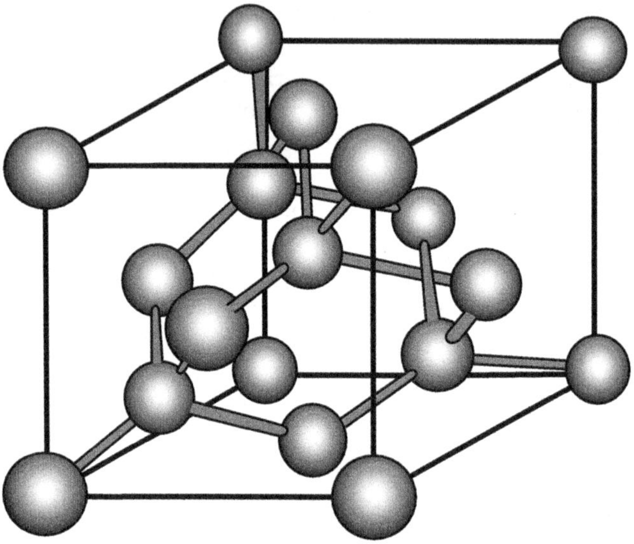

FIGURE 5.5
Silicon crystal structure.

On the contrary, {110} planes exhibit highest etching rate [8–10] in the alkaline etchant due to its low packing density or the orientation of the bonds. Low packing density facilitates the longer channel to propagate water molecule in that plane (Figure 5.5).

Incidentally, this {110} is the tangent plane on the intersecting {111} planes, which is the convex corner of the structure, which means the convex corner lies on the {110} plane. Due to the high etch rate as discussed, the undercutting is observed at the convex corners. The convex corners with just one dangling bond have been described elaborately by Pal et al. [13] and are shown in Figure 5.6.

For convex corners, the fastest etching planes dominate the 3D shape.

As we have discussed earlier, {110} plane appeared at the intersection of two {111} planes. Due to the high etching rate of the {110} plane, corner undercutting occurs, and during this etching, some more high-index planes such as {311}, {411}, {331}, {212}, {772} appear, which are shown in the following figure.

This undercutting is a function of etch time and thus directly related to the desired etch depth. An undercut ratio is defined as the ratio of undercut to etch depth.

5.2.1.2.1 Compensation Structures

Undercutting can also be reduced or even prevented by corner compensation structures that are added to the corners in the mask layout. Depending on the etching solutions, different corner compensation schemes are used (Figure 5.7).

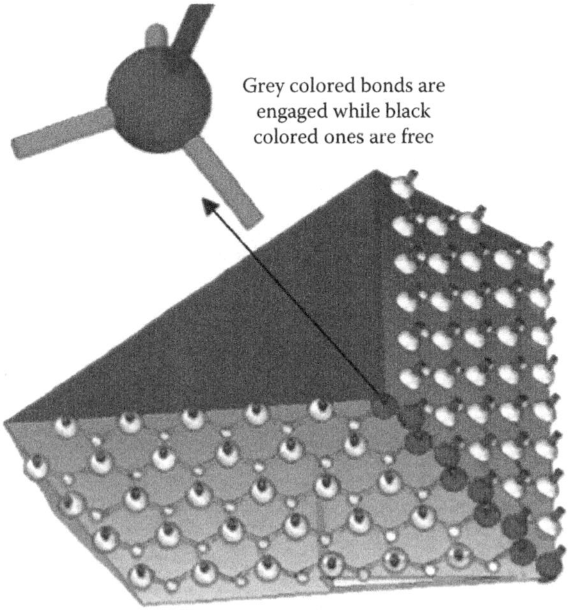

FIGURE 5.6
Convex corner with dangling bond. (From Pal, P. and Singh, S.S., *Micro Nano Syst. Lett.*, 2013.)

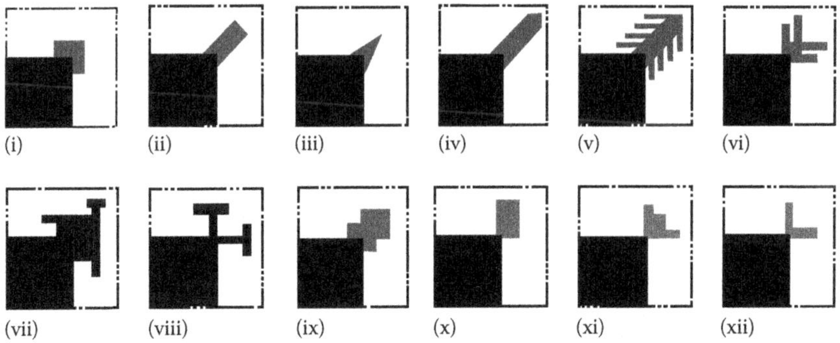

FIGURE 5.7
Different compensation structures. (From Abu-Zeid, M., *J. Electrochem. Soc.*, 131, 2138, 1984; Wu, X.P. and Ko, W.H., *Sens. Actuators A*, 18, 207, 1989; Puers, B. and Sansen, W., *Sens. Actuators A*, 23, 1036, 1990; Mayer, G.K. et al., *J. Electrochem. Soc.*, 137, 3947, 1990; Offereins, H.L. et al., *Sens. Actuators A*, 25, 9, 1991; Bao, M. et al., *Sens. Actuators A*, 37, 727, 1993; Scheibe, C. and Obermeier, E., *J. Micromech. Microeng.*, 5, 109, 1995; Zhang, Q. et al., *Sens. Actuators A*, 56, 251, 1996; Enoksson, P., *J. Micromech. Microeng.*, 7, 141, 1997; Fan, W. and Zhang, D., *J. Micromech. Microeng.*, 16, 1951, 2006; Pal, P. et al., *J. Micromech. Microeng.*, 17, 110, 2007.)

5.2.2 Dry Etching

This technology is expensive compared to wet etching. But if the target material is thin film and intended to create high-resolution thin-film structures or if higher-depth vertical sidewalls are required, one should go through dry etching. Dry etching is an advanced technology compromising the expenditure of fabrication.

Dry isotropic etching is performed in plasma, which chemically attacks the materials being etched; in the case of silicon, the usual gas is SF_6.

If a DC electric field is created between the top and the bottom of the plasma chamber, the ions attain a drift velocity with a mean free path and move in the vertical direction and attack the substrate physically in a highly directional manner. In this fashion, the vertical side walls are formed. But practically, the walls always have some curvature in their sides.

For the patterning of thin films instead of anisotropic processes, reactive ion etching (RIE) is usually preferred as the film being etched will have better vertical edges as presented by the mask material. RIE is already discussed in an earlier chapter.

It should be noted that wet etching has the advantage of requiring less equipment, and it also demonstrates higher selectivity between different materials.

5.3 Surface Micromachining

As the name suggests, surface micromachining processes are responsible for creating microstructures that exist near the surfaces of a substrate. Bulk micromachining creates devices by etching of the wafer and therefore termed as subtractive process, whereas surface micromachining builds devices on top of the wafer, maybe in multilayer, and can be termed as additive process. Surface micromachining is used to fabricate mechanical features such as cantilevers, membranes or free-moving structures on top of a wafer surface.

5.3.1 Processes

A usual surface micromachining process consists of depositing thin films on a wafer, photo patterning the films and then etching the films to generate the pattern. Surface micromachining is done to create moving functional mechanical parts on top of the substrate, that is why an alternating layer of two or more different materials is to be deposited on the substrate. Each additional layer increases the level of complexity and a resulting difficulty in fabrication. The layer that is to be removed from the wafer surface to create a floating structure is called the sacrificial layer; generally, photoresist is used. The structural material out of which the free-standing structure is made (generally, polycrystalline silicon or polysilicon, silicon nitride and aluminium) will form the mechanical elements, and after removal, the sacrificial layer creates the voids between the mechanical element and the wafer. If the sacrificial layer is SiO_2, then HF is used for its removal. In the same way, if it is photoresist, then the *release* process is performed by acetone. This phenomenon is called lift-off (Figure 5.8).

Much more fabrication steps are required in surface micromachining than bulk micromachining and hence are more expensive. Where extreme sophisticated mechanical structures are required, surface micromachining is suitable for that application. Surface micromachining could also be performed using dry etching methods.

Plasma etching of the silicon substrate with SF_6/O_2-based and CF_4/H_2-based gas mixtures is advantageous because high selectivity for photoresist, silicon dioxide and aluminium masks can be achieved.

Surface micromachining cannot be a successful process until the sacrificial layers are removed completely from the surface to free the structural elements so that it can be actuated. The yield of the fabrication related to surface micromachining is subjected to a phenomenon called stiction. Stiction refers to the sticking of structural elements with the substrate or the neighbouring elements.

It was investigated that capillary forces arise from the solvent, or the unintentional electrostatic and van der Waals forces are responsible for producing this permanent sticking problem after the device is dehydrated.

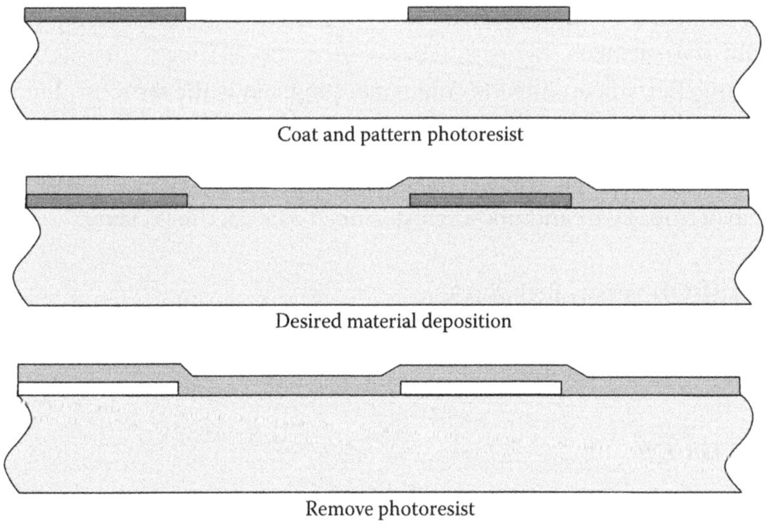

FIGURE 5.8
Schematic diagram of surface micromachining.

5.3.2 Hurdles

Surface micromachining demands the structural materials, sacrificial materials and chemical etchants simultaneously. The structural materials must not only have the desired physical and chemical properties but it must also have sufficient mechanical strength as during the lift-off process, high fracture stresses and good wear resistance are highly required to circumvent the device failure during fabrication. The sacrificial materials must have good adhesion property and low residual stresses to fabricate the floating structure properly. The viscosity of the etchant should be checked properly; otherwise, some residue would always exist to make the effort in vain.

Another important point to be monitored is that the etchant to remove the sacrificial layer must have excellent etch selectivity so that it would not damage the structural material in any way.

The common materials used in surface micromachining are as follows:

1. Poly-Si/silicon dioxide conjugate; poly-Si is used as the structural material and LPCVD-deposited oxide is the sacrificial material. HF solution is used for removing the oxide layer without affecting the poly-Si layer.
2. Polyimide/aluminium conjugate; here, the polyimide is the structural material and aluminium is the sacrificial material. Acidic solution is used to eliminate the aluminium sacrificial layer.
3. Silicon nitride/poly-Si conjugate; silicon nitride is used as the structural material, whereas poly-Si is the sacrificial material. Silicon anisotropic etchants such as KOH and TMAH are used to eradicate the poly-Si layer.
4. Tungsten/silicon dioxide conjugate; tungsten is the structural material with oxide as the sacrificial material. HF solution is used to remove the oxide layer.
5. Metallic layer/photoresist conjugate; the metallic layer behaves as a structural layer and photoresist is used as a sacrificial layer.

5.3.3 Lift-Off versus Etch Back

Etching followed by lithographic technique can be done in two ways that are discussed in the following.

5.3.3.1 Lift-Off

This follows the following steps (Figure 5.9):

- First photoresist is coated on the substrate.
- Patterning is done on the photoresist layer by normal lithographic technique.

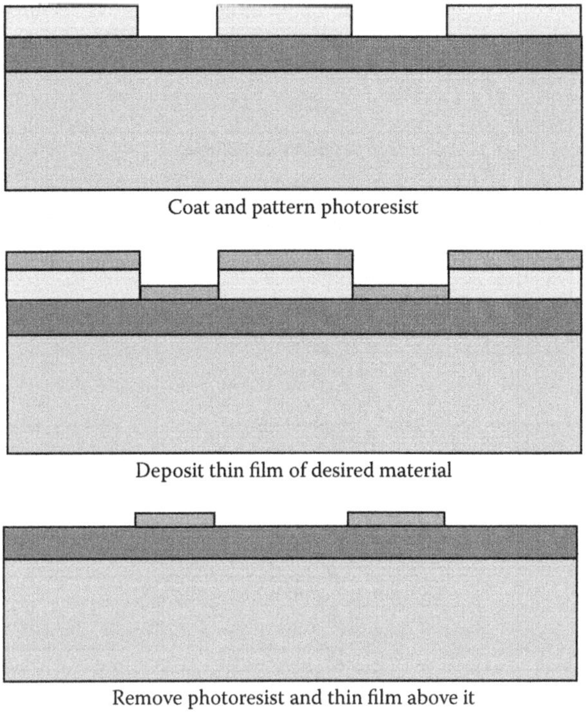

FIGURE 5.9
Lift-off procedure.

- Thereafter, metals are deposited on the patterned photoresist layer.
- Photoresist is removed. Unwanted material is lifted off when resist is removed.
- Free-floating patterns are formed.
- Photoresist has the opposite polarity of the final film.

5.3.3.2 Etch Back

This follows the following steps (Figure 5.10):

- Deposit the metal layer first.
- Photoresist is applied at the top of the layer to be patterned.
- Normal lithography is done, and the unwanted portion of photoresist is removed by organic developer.
- Pattern is formed. Unwanted metal is etched away by acidic solution.
- Finally, patterned photoresist is washed out.

Photoresist has the same polarity of the final film.

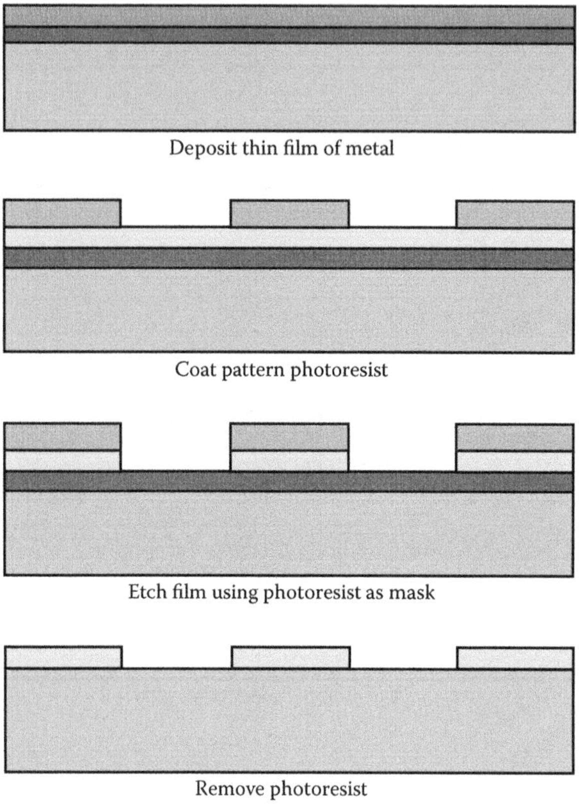

FIGURE 5.10
Etch-back procedure.

5.4 Etch-Stop Technique

Etch-stop technique is very much useful while fabricating a thin membrane as it allows the termination of the etching process at a predetermined depth. Broadly, etch-stop technique can be classified into two groups. One is intrinsic etch stop and the other is extrinsic etch stop. In intrinsic etch stop, no external power source is required, and it is totally dependent on the chemicals used, whereas in extrinsic etch stop, external power source is required. Any type of illumination may serve the purpose as it provides the required activation energy to start the reaction. Extrinsic etch-stop technique is not preferred so much as the accuracy is low.

Further intrinsic etch-stop technique can be subdivided into three classes.

1. Time etch stop where the etching is stopped by taking the wafer out of the solution. This process is not reliable as one has to check the etch depth repeatedly by taking the wafer out of the solution. Only rough estimation of etch depth is possible with a bumpy surface finish.
2. The second one is boron etch stop.
3. The third one is etch stop using underlying masking layer.

Among them, boron etch stop is mostly preferred as the precise thickness of the membrane is possible using this technique.

5.4.1 Boron Etch Stop

It should be kept in mind that etching process is basically a charge-transfer mechanism; therefore, it is understood that there is an effect of dopant on the etch rates. As the density of the charge will vary in the doped region with respect to the undoped region, dopant type and dopant concentration play a key role in the charge-transfer process. The schematic of the set-up is shown in Figure 5.11. In particular, highly doped materials might have higher etch rate characteristics than lightly doped silicons depending on the reactions involved between the material and the solution. For example, if the etching solution itself is a source of holes (generated by the reaction) as well as hydroxyl groups, and

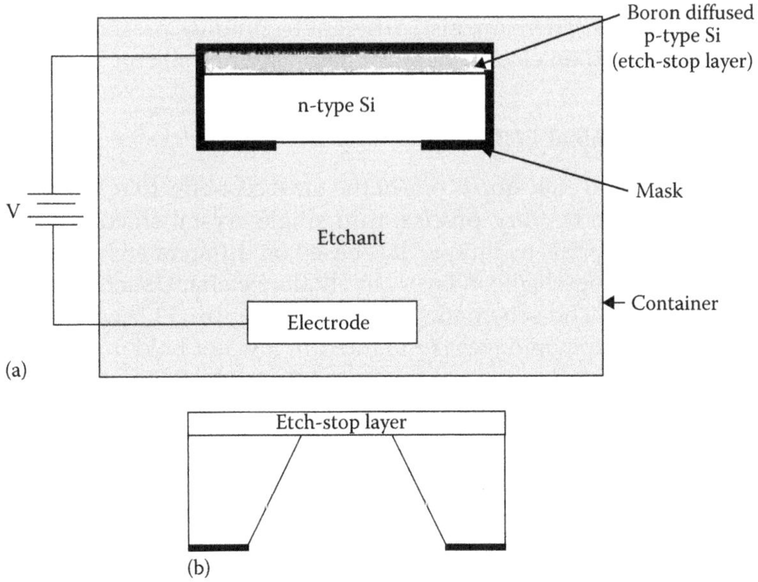

FIGURE 5.11
(a) Schematic of boron etch-stop set-up and (b) resultant structure.

also contain an agent whose reacted species is soluble in the etchant solution, these holes obviously elevate the oxidation state of the silicon; on the other hand, OH groups are necessary to oxidize the silicon (Figure 5.11).

As described by Peterson [25], in the HNA ($HF+HNO_3+H_2O$) system, both the holes and the hydroxyl groups are generated by the strong oxidizing agent HNO_3, while fluorine from HF forms the soluble species H_2SiF_6. The overall reaction is autocatalytic since HNO_3 plus trace impurities of HNO_2 combine to form additional HNO_2 molecules.

$$HNO_2+HNO_3+H_2O \rightarrow 2HNO_2+2OH^-+2h^+$$

But if the etchant is anisotropic such as EDP, TMAH or KOH, etching rate decreases drastically, and ultimately it goes to zero if the samples are heavily doped with boron (~10^{20} cm^{-3}). The high concentrations correspond to an average separation between boron atoms of 20–25 A°, where the solid solubility limit of boron is 5×10^{19} cm^{-3}. At high concentrations, the interstitial sites of silicon are occupied by boron atom, and the strong B–Si bond tends to bind the lattice more firmly. Therefore, the amount of energy required to remove a silicon atom from the bulk is impossible to get, and that is why etching stops. Moreover, high concentrations of boron, converted to boron oxides and hydroxides in an intermediate chemical reaction, would passivate the surface of silicon and prevent further detachment of silicon atom.

In the extrinsic etch-stop technique, external power is required to prevent the etching process. There are also different techniques present. These techniques are all based on electrochemical passivation of silicon.

5.4.2 Electrochemical Etch Stop

The electrochemical etch stop is one of the most cost-effective and most commonly used methods. Very precise, thin single-crystal silicon membranes can be produced by this technique. It is based on different etching potentials of n- and p-type doped silicon layers in alkaline etchant, such as potassium hydroxide (KOH). The schematic is shown in Figure 5.12, where a p-type wafer with an n-type epi-layer is mounted in a wafer holder. The n-type Si is biased positive with respect to a counter electrode. Etching of the p-type silicon stops at the pn junction. Table 5.1 summarizes the characteristic of different etchants.

This was first proposed by Waggener (1970). The structure usually consists of a p-type silicon wafer deposited on an n-type epitaxial layer. There is a counter electrode to apply the potential; generally, platinum is used. The epitaxial layer is connected to the positive terminal of the battery, and platinum counter electrode is connected to the negative terminal of the battery. The whole experiment is carried in a container containing alkaline solution. The metallic contact into the n-type epi-layer is protected from the corrosive etchant by

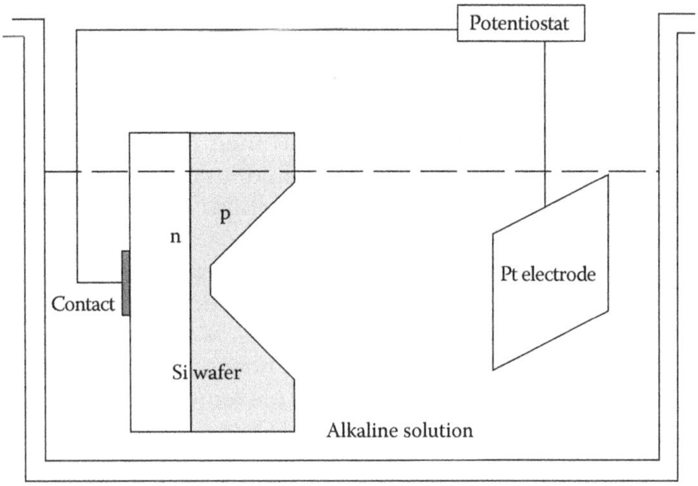

FIGURE 5.12
The set-up for the electrochemical etch-stop technique.

TABLE 5.1
Summary of Different Approaches Using Different Etchants

Different Wet Chemical Approaches	Effect on SiO_2	Pattern Formation	Etched Bottom Layer	Possibility of Thin Membrane Formation
Conventional KOH–IPA	Effect on oxide layer	Well-defined pattern	Little bit textured.	Pyramidal hillock observed on the etched surface.
Hydrazine–H_2O	Nil	Well-defined pattern	More textured.	More hole formation on bottom layer.
Hydrazine–NaOH	Oxide layer etched but slow compared to conventional approach	Well-defined pattern but more time required	Little bit textured.	Hole formation on etched membrane.
EDP anisotropic etching	Nil	Well-defined pattern	Flat-etched surface and no texturization.	No hole formation on the surface. Very thin membrane can be produced.
TMAH anisotropic etching	Less effect on oxide layer	Well-defined pattern	Little bit textured. Better than KOH.	Very few holes observed.

placing it in a wafer holder. The pn junction is reverse biased. The alkaline solution etches the p layer anisotropically as discussed earlier, and the p-type bulk floats in the solution at approximately the same potential as the solution.

If the applied bias is sufficiently high and the silicon surface acquires a positive potential with respect to the passivation potential, an anodic current flows to the silicon/solution interface and an oxide layer grows on the surface by the process of oxidation and reduction. The silicon left behind is now passivated, and gradually, the etching stops. Prior to the experiment, these criteria should be met first: the breakdown voltage of the pn junction should be high and the reverse leakage current should be lower than the Si passivation current, and for this, the pn junction is of good quality.

Furthermore, most of the potential drop is now across the depletion region of the pn junction. The p layer is in direct contact with the etching solution, and the solution etches it chemically as it is isolated from the positive terminal of the power supply. In this way, when the etch front reaches the pn junction, it is destroyed locally.

5.4.3 Photo-Assisted Electrochemical Etch Stop (for n-Type Silicon)

This etch-stop technique requires an external energy in the form of a high-intensity light source. Etching stops on a p-type epitaxial layer. A platinum/titanium film is sputter deposited on the n-type bulk and patterned accordingly to take out the contact. The set-up is illustrated in Figure 5.13, where the p-type layer is protected from the etchant by a nitride layer. The n-type substrate is contacted with a platinum film. When the etching stops, the reaction indicated in the figure occurs and the reverse current flows through the pn junction.

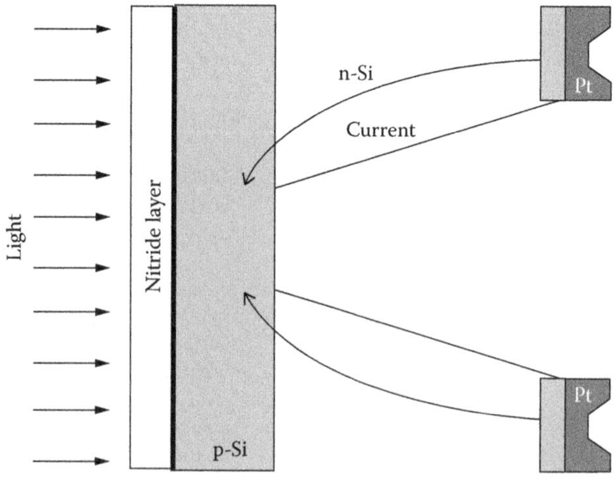

FIGURE 5.13
Device structure used for the photovoltaic etch-stop technique.

A silicon nitride layer is deposited on the p-type epitaxial layer to protect the epi-layer from the solution. As an etchant, KOH solution is used. When a strong source of light illuminates the substrate, the etching process stops at the epi-layer. The platinum film and the n-type silicon substrate acted galvanically.

The reactions involved with this technique are as follows:

$$Si + 4OH^- + 4h^+ \rightarrow SiO_2 + 2H_2O$$

$$O_2 + 2H_2O + 4e^- \rightarrow 4OH^-$$

The actual reason behind this etch stop is stated as follows: the light energy breaks the covalent bonds and generates the electron/hole pairs. These photogenerated carriers are lost by recombination. In the bulk of the silicon, these recombine, while in the depletion layer, these are separated by the electrical field. Holes driven into the p-type epi-layer cannot react as long as this layer is protected by the n-type bulk. Note that at the front, the epi-layer is covered with a nitride layer. A photopotential develops and the photogenerated carriers are lost by recombination. Electron/hole pairs created in the n-type substrate are also very likely to recombine; kinetics of surface reactions involving electrons and holes is slow compared to recombination kinetics. An anodic current flow, required for the passivation of silicon, is not present in the structure. The n-type substrate is therefore chemically etched. A galvanic effect might be expected due to oxygen reduction at the platinum. However, since the areas of n-Si and Pt exposed to the solution are comparable and the passivation current density in KOH solutions is high, galvanic passivation of the n-type Si is unlikely. Once the p-type layer is exposed to the solution, photogenerated holes separated by the electric field of the depletion layer are free to react at the p-type semiconductor/solution interface. The silicon surface is oxidized. The photogenerated electrons can react at the n-type side walls or more likely at the platinum/solution interface, where hydrogen is generated. Now, there is an internal anodic photocurrent. If the photocurrent is sufficiently high, etching of the p-type epi-layer is prevented. There are two important requirements for the etch stop to work. First, there must be a high-power light source proving that an anodic photocurrent is essential. Second, there must be a sufficiently large platinum electrode. If this is not the case, the photocurrent will be limited by the reduction of water to hydrogen gas at the platinum/solution interface. The reduction of water at an n-type silicon interface is less efficient.

This process requires a strong light source to activate, which makes it a little bit complicated.

5.4.4 Etch Stop at Thin Films: Silicon on Insulator

A basic silicon-on-insulator (SOI) structure consists of a thin silicon film stuck over an insulating layer. The insulating layer, which is called buried layer, is

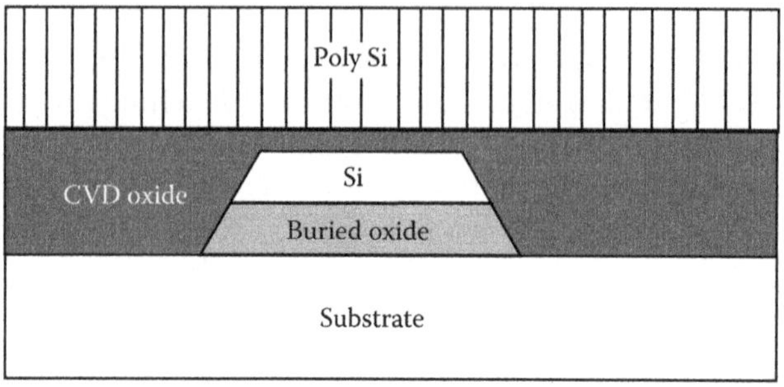

FIGURE 5.14
Etch-stop technique on silicon-on-insulator.

deposited on the top of a bulk silicon substrate as shown in Figure 5.14. This buried layer acts as an etch mask layer. Materials that can be used are silicon oxide, silicon nitride, silicon carbide, etc. The top silicon layer is used for active devices, while the bottom bulk silicon substrate acts as a mechanical support. Because the sandwiched insulating film is normally silicon dioxide, it is also known as buried oxide.

There are various methods to fabricate these buried layers. The first one is the implanted reactive ions that react with the bulk silicon to form the buried etch-stop layer. Oxygen may be implanted to form silicon oxide, nitrogen to form silicon nitride and carbon to form silicon carbide, though the thickness of the structural layer is restricted. SOI has long been the foremost latest development in the CMOS technology.

5.5 High-Aspect-Ratio Micromachining

High-aspect-ratio micromachining (HARM) is a process that involves micromachining as a tooling step followed by injection moulding or embossing and, if required, by electroforming to replicate microstructures in metal from moulded parts. It is one of the most attractive technologies for replicating microstructures at a high-performance-to-cost ratio and includes techniques known as LIGA. Products micromachined with this technique include high-aspect-ratio fluidic structures such as moulded nozzle plates for inkjet printing and microchannel plates for disposable microtitre plates in medical diagnostic applications. The materials that can be used are electroformable metals and plastics, including acrylate, polycarbonate, polyimide and styrene. One example of this technique is given in Figure 5.15.

FIGURE 5.15
Scanning electron microscope image of 100 μm deep, 1 μm diameter basic silicon-on-insulator gyroscope (HARM technique).

5.5.1 LIGA

LIGA is a German acronym for Lithographie, Galvanoformung, Abformung (Lithography, Electroplating and Moulding) and is used to form high-aspect-ratio structures.

LIGA is an important tooling and replication method for high-aspect-ratio microstructures [13]. The flow chart is shown in Figure 5.16. The technique employs x-ray synchrotron radiation to expose thick acrylic resist of polymethylmethacrylate (PMMA) under a lithographic mask. The exposed areas are chemically dissolved, and in areas where the material is removed, metal is electroformed, thereby defining the tool insert for the succeeding moulding step. LIGA is capable of creating very finely defined microstructures up to 1000 μm high. LIGA is limited by the need to have access to an x-ray synchrotron facility. A compromise that combines some features of LIGA with surface micromachining eliminating the need for exposure to x-rays has been developed and is known as SLIGA (sacrificial LIGA). It replaces the thick PMMA photoresist with polyimide as the electroplating mould, thus enabling compatible conventional IC batch processing. HARM production methods have provided radically new ways to produce micromachined parts for MEMS devices at relatively low cost. In particular, techniques such as SLIGA enable the production of MEMS components with much lower manufacturing infrastructures in terms of investment, facilities and access to advanced materials and technology.

Other microreplication techniques can be combined to generate a preform for the tool insert.

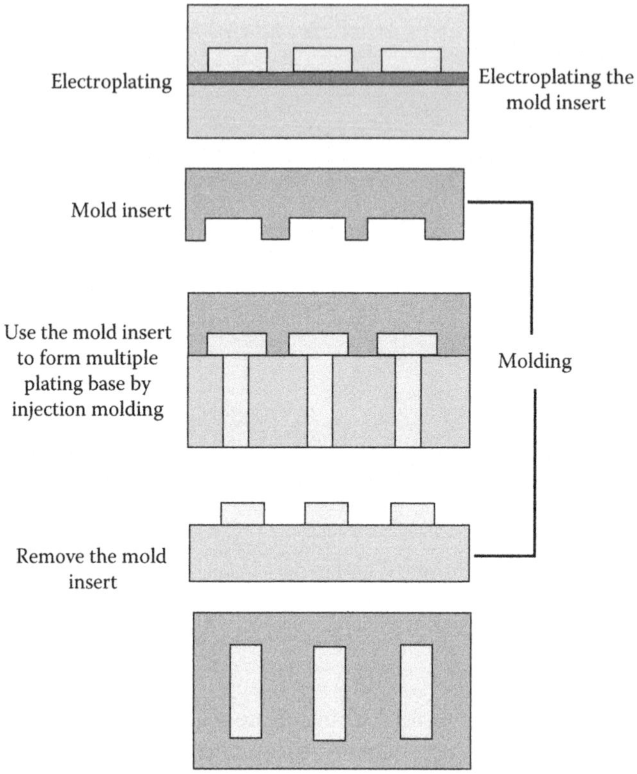

FIGURE 5.16
Flow chart of LIGA.

These include laser ablation, ultraviolet lithography and mechanical micromachining, which includes electric discharge machining (EDM) and diamond milling. EDM is a relatively new approach that uses machine shop production techniques and offers the capability to make parts out of most conductive materials. Unfortunately, as a spark erosion technique, it is slow and not ideal for batch processing but has found many applications for MEMS prototype production.

5.5.2 Laser Micromachining

Laser micromachining processes are not parallel and hence not fast enough for effective MEMS fabrication. Nonetheless, they have utility in specialty micromachining or making moulds. Excimer laser micromachining is used particularly for the micromachining of organic materials (plastics, polymers, etc.) as the material is not removed by burning or vaporization. Hence, material adjacent to the machined area is not melted or distorted by heating effects. The schematic of the set-up is shown in Figure 5.17.

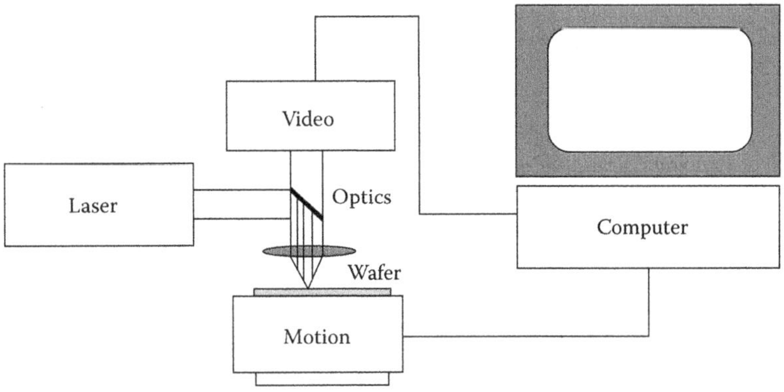

FIGURE 5.17
Laser micromachining set-up.

References

1. T. Suzuki, K. Kunihara, M. Kobayashi, S. Tabata, K. Higaki and H. Ohnishi, A micromachined gas sensor based on a catalytic thick film/SnO_2 thin film bilayer and thin film heater, *Sens. Actuators B* 109, 185–192 (2005).
2. S. Ray, P. S. Gupta and G. Singh, Electrical and optical properties of sol-gel prepared Pd-doped SnO_2 thin films: Effect of multiple layers and its use as room temperature methane gas sensor, *J. Ovonic Res.* 6, 23–34 (2010).
3. F. Becker, Ch. Krummel, A. Freiling, M. Fleischer and C. Kohl, Decomposition of methane on polycrystalline thick films of Ga_2O_3 investigated by thermal desorption spectroscopy with a mass spectrometer, *J. Anal. Chem.* 358, 187–189 (1997).
4. X. Tang, J. Hao and J. Lim, Advances in water resources and hydraulic engineering: Proceedings of 16th IAHR-APD congress and 3rd Symposium of IAHR-ISHS, China, *Front. Environ. Sci. Eng.* 3, 265 (2009).
5. A. Rothschild and Y. Komem, The effect of grain size on the sensitivity of nanocrystalline metal-oxide gas sensors, *J. Appl. Phys.* 95, 6374–6380 (2004).
6. E. Comini, C. Baratto, G. Faglia, M. Ferroni, A. Vomiero, and G. Sberveglieri, Quasi-onedimensional metal oxide semiconductors: Preparation, characterization and application as chemical sensors, *Prog. Mater. Sci.* 54, 1–67 (2009).
7. J. Min and A. J. Baeumner, Characterization and optimization of interdigitated ultramicroelectrode arrays as electrochemical biosensor transducers, *Electroanalysis* 16, 724–729 (2004).
8. K. Furuya, K. Nakanishi, R. Takei, E. Omoda, M. Suzuki, M. Okano, T. Kamei, M. Mori and Y. Sakakibara, Nanometer-scale thickness control of amorphous silicon using isotropic wet-etching and low loss wire waveguide fabrication with the etched material, *Appl. Phys. Lett.* 100, 251108 (2012).
9. K. Sato, M. Shikida, T. Yamashiro, M. Tsunekawa and S. Ito, Roughening of single-crystal silicon surface etched by KOH water, *Sens. Actuators A* 73, 122–130 (1999).

10. M. Shikida, K. Sato, K. Tokoro and D. Uchikawa, Differences in anisotropic etching properties of KOH and TMAH solutions, *Sens. Actuators A* 80, 179–188 (2000).
11. I. Barycka and I. Zubel, Silicon anisotropic etching in alkali solutions: I, *Sens. Actuators A* 70, 250–259 (1998).
12. H. Schroder, E. Obermeier, A. Horn and G. K. M. Wachutka, Convex corner undercutting of {100} silicon in anisotropic KOH etching: The new step-flow model of 3-D structuring and first simulation results, *J. Microelectromechanical Syst.* 10, 88–97 (2001).
13. P. Pal and S. S. Singh, A simple and robust model to explain convex corner undercutting in wet bulk micromachining, *Micro Nano Syst. Lett.* 1, 1–6 (2013).
14. M. Abu-Zeid, Corner undercutting in anisotropically etched isolation contours, *J. Electrochem. Soc.* 131, 2138–2142 (1984).
15. X. P. Wu and W. H. Ko, Compensating corner undercutting in anisotropic etching of (100) silicon, *Sens. Actuators A* 18, 207–215 (1989).
16. B. Puers and W. Sansen, Compensation structures for convex corner micromachining in silicon, *Sens. Actuators A* 23, 1036–1041 (1990).
17. G. K. Mayer, H. L. Offereins, H. Sandmaier and K. Kuhl, Fabrication of non-underetched convex corners in anisotropic etching of (100) silicon in aqueous KOH with respect to novel micromechanic elements, *J. Electrochem. Soc.* 137, 3947–3951 (1990).
18. H. L. Offereins, K. Kühl and H. Sandmaier, Methods for the fabrication of convex corners in anisotropic etching of (100) silicon in aqueous KOH, *Sens. Actuators A* 25, 9–13 (1991).
19. M. Bao, Chr. Burrer, J. Esteve, J. Bausells and S. Marco, Etching front control of <110> strips for corner compensation, *Sens. Actuators A* 37–38, 727–732 (1993).
20. C. Scheibe and E. Obermeier, Compensating corner undercutting in anisotropic etching of (100) silicon for chip separation, *J. Micromech. Microeng.* 5, 109–111 (1995).
21. Q. Zhang, L. Liu and Z. Li, A new approach to convex corner compensation for anisotropic etching of (100) Si in KOH, *Sens. Actuators A*, 56, 251–254 (1996).
22. P. Enoksson, New structure for corner compensation in anisotropic KOH etching, *J. Micromech. Microeng.* 7, 141–144 (1997).
23. W. Fan and D. Zhang, A simple approach to convex corner compensation in anisotropic KOH etching on a (100) silicon wafer, *J. Micromech. Microeng.* 16, 1951–1957 (2006).
24. P. Pal, K. Sato and S. Chandra, Fabrication techniques of convex corners in a (100)-silicon wafer using bulk micromachining: A review, *J. Micromech. Microeng.* 17, 110–132 (2007).
25. Kurt E. Petersen, Silicon as a mechanical material, *Proc. IEEE* 70, 420–457 (1982).
26. M. McNie, D. King, C. Vizard, A. Holmes, K. W. Lee, High aspect ratio micromachining (HARM) technologies for microinertial devices, *Microsyst. Technol.* 6, 184–188 (2000).

6

Microheaters for Gas Sensor

6.1 Introduction

A gas sensor can be defined as a device that identifies and confers the information about the surrounding gas atmosphere, which comprises many types of hazardous and toxic gases. The device mainly consists of a sensing layer on top of the substrate and a signal-conditioning unit. There has been extensive research on gas sensors during the last few years, particularly based on either the surface characteristics or the bulk electrolytic properties of ceramics [1–3]. Nanocrystalline metal oxide–based solid-state sensors [4–7] were found to be promising among the sensor family, as they detect a wide variety of gases like O_2, H_2, CO and NO_2 and different volatile organic compounds [8–11] like propane, methane, ethanol, methanol and acetone with a low operating temperature, small size, low cost, high sensitivity and relatively simple associated circuitry particularly in domestic and industrial environments.

A solid-state sensor can be used to detect various gases of multiple ranges by varying the sensor operating temperature [12]. Multiple ranges are needed to identify the level of toxic concentrations as well as explosive concentrations simultaneously. The solid-state sensor is able to detect gas in both the ranges. This greatly simplifies the sensor system design because it minimizes the use of complicated multiple sensor technologies that must be designed and synthesized in a different way.

Temperature is one of the main factors that determine the sensitivity, selectivity and response time of sensors. So as far as temperature is concerned, one of the key components of a chemical gas sensor is a microelectromechanical system (MEMS) microheater.

High operating temperature (≥300°C) is another drawback for most of the gas sensor systems (Taguchi type) because this ultimately results in high power consumption. Low power consumption is a fundamental requirement for a sensor system with an acceptable battery lifetime especially in field application. However, the application of silicon MEMS technology may permit the desired benefits of reduced thermal mass, miniaturization, low power, reproducibility and low unit cost. The necessity of sensors with improved performances leads to continuous research efforts aiming at the optimization of sensing materials, of device design and of the operating modes.

For the semiconductor gas sensors, an elevated temperature with uniform temperature distribution throughout the sensing layer is a necessary requirement as it often enhances the sensitivity of the sensor. For this reason, the MEMS microheater [13–16] is one of the key components for the chemical gas sensors in raising the required temperature with temperature uniformity. The temperature uniformity depends mainly on the membrane materials and on the geometry of the microheater. Recent trends on application of metal-oxide gas sensors request for thermal as well as mechanical performance such as good mechanical stability at high temperatures [17–20]. For these objectives, the thermal characteristics of the microhotplate have to be well known and optimized, mainly with respect to power consumption, transient response and uniform temperature distribution, by controlling the heat losses, dielectric materials and heater configuration.

Screen-printed ceramic gas sensors need to be developed in power consumption, mounting technology and selectivity aspects. The power consumption of screen-printed devices is in the range of 1–5 W [21], which is unacceptable for application as it allows only the use of battery-driven elements. The mounting of the overall hot ceramic element is difficult. One has to find such designs that ensure good thermal insulation between sensor element and housing as well as high mechanical stability. Good thermal isolation is thereby needed not only to minimize the overall power consumption but also to enable the integration of signal processing electronics in the same housing. Sufficient selectivity of metal oxide sensors can up to now be achieved only if the sensor is used in an application where the number of gases is limited such that cross sensitivities can be neglected or if several sensors are put together to an array. Even though the use of array is very promising with respect to sensor selectivity, one has to have in mind that the use of sensor arrays leads at the same time either to an increased size of the sensor element or to the usage of several separate sensor elements and thus to an increased power consumption.

Most solid-state sensors have four or six pins, depending on how the microheater and output electrodes are connected. In this chapter, microheater design issues and fabrication are discussed. Different types of heater element, structures and geometries are analyzed in order to obtain an optimal design of metal oxide–based gas sensor.

6.2 Need of Microheater

The number of applications for MEMS microheater devices is increasing rapidly, as they are the key components in a gas sensor system including humidity as well as toxic gas sensors. Microheater provides the requisite temperature for metal oxide–based gas sensor. These temperatures ultimately pursue various factors such as the conductance of the sensing layer and the capacity of a particular gas absorbed and provide the activation energy to necessitate the reaction rates between different absorbates on the sensing layer surface. Hence, a microheater plays a decisive role for the procurement of signals from the gas sensor surface.

A microhotplate is nothing but a miniaturized thin membrane that is thermally isolated from the bulk counterparts, which generally consists of a microheater to heat up the sensing layer and a temperature sensor to detect the temperature raised and contact electrode and to take the output of the sensor element. These types of MEMS structures are really useful while designing a power-efficient device. Controlled-temperature microhotplate platforms have, therefore, explored a new area in chemical sensor research.

Microhotplate is a generic structure in MEMS-based gas sensor system that can be used by coating with different sensing films, either by drop-coating or by spin-coating techniques. These coatings are generally metal oxide nanocomposites that are able to detect hazardous gases like carbon monoxide, nitrogen dioxide or ammonia or combustible gases like LPG, hydrogen and methane.

Since 1990s, micromachining technology including surface and bulk micromachining of silicon has led to the expansion of miniaturized hotplate structures with very low power consumption (50 mW) and short thermal response (<10 ms).

6.3 Types of Microheater

The function of a microheater is to raise the temperature to a predetermined level. The sensing layer should get heated and should detect the toxic gases from the ambient. Obviously, this sensing layer occupies a region called the active region of the device, which is the heart of a sensor system. The main objective is to maintain a uniform temperature throughout the active region. To achieve this, different device structures have been adopted. The membrane can be devised either through front-side etching or through backside etching; wafer or sacrificial layer or both can be used in the process. Front-side etching results in a suspended-membrane type, and backside etching that removes bulk counterparts is called a closed-membrane type.

Researchers have developed the microhotplate structure using different metals and metal alloys and also using different heater structures. Whatever may be the issue, the key aspect is that the heater material should sustain the high temperature without damage with low thermal expansion and the membrane should be a good dielectric with low thermal conductivity. There are two types of membranes: closed membrane and suspended membrane.

6.3.1 Closed-Membrane Type

Previously, metal oxide gas sensors used suspended-type microheaters on the SiO_2/Si_3N_4 composite thin membrane as their operating temperature was very high, though this results in thermal stress causing the generation of microcracks, which in turn leads to shorter lifetime of the device. Moreover, the use of composite membrane underneath the microheater doesn't in anyway help to increase the temperature uniformity of the device due to low thermal conductivity of the dielectric. This nonuniformity further leads to a reduced sensing area, which degrades the sensitivity of the gas sensor. Moreover, they are relatively less stable at higher temperatures. In recent years, the use of nanomaterials as the sensing layer has been a motivating area of research for their enormously improved surface-to-volume ratio compared to the bulk counterpart. This leads to an opportunity to lower the operating temperature of the metal oxide semiconductor. This low operating temperature facilitates the use of closed membrane in place of suspended membrane to increase the long-term stability of the device.

In the closed-membrane type, there is a silicon membrane underneath the heater area, which serves as the heat distributor among the membrane and also stabilizes it mechanically. Generally, in the closed-membrane type, a stack of SiO_2/Si_3N_4 dielectric layer on top of the thin silicon layer has been adopted to make it robust and durable in terms of long-term reliability.

Simon et al. [22] fabricated the closed-membrane-type sensor for the first time by wet etching technique. Many researchers utilize silicon dioxide/silicon nitride as membrane and as insulator with a typical thickness of 1–2 μm [23–25]. Silicon dioxide and silicon nitride possess low thermal conductivities and can thus provide good thermal isolation between the heated active area and the membrane rim. Silicon nitride layers generally have large tensile stress, and silicon oxide layers are compressive. The mechanical properties (Young's modulus E and mean residual stress σ_0, Poisson's ratio υ) of Si_3N_4 thin films (about 100 nm in thickness) and Si_3N_4/SiO_2 bilayers (about 200 nm in thickness) have been investigated by many researchers [26,27]. A thin-film gas sensor is utilized on a low power consumption micromachined silicon structure.

The use of SiO_2/ Si_3N_4 composite membrane minimizes the mechanical stress developed due to high temperature by utilizing the dual effect of compressive stress of SiO_2 and tensile stress of Si_3N_4. Ni et al. [28] have demonstrated the effective means to microfabricate silicon dioxide and silicon nitride membranes with varying sizes from small membranes of 200 μm × 200 μm to significantly larger membranes of 9000 μm × 9000 μm. Benn et al. [29] use silicon carbide as microhotplate for high-temperature chemical sensing.

Xiang et al. [30] reported the mechanical behaviour of Cu thin film through the bulge test method. The insulating layer was SiN_x LPCVD/SiN_x PECVD

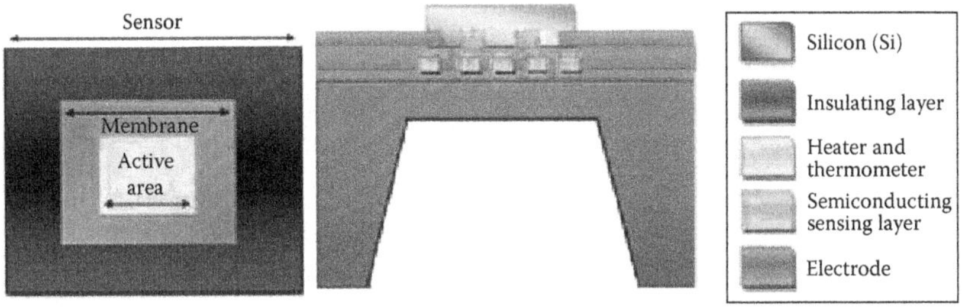

FIGURE 6.1
Schematic of a closed-membrane-type gas sensor. (Adapted from Roy, S. et al., *Sens. Lett.*, 9, 1382, 2011.)

rectangular (2.4 mm x 10 mm) bilayer membranes of about 100 nm in thickness. Tripp et al. [31] have reported results on 100 nm thick Al_2O_3 circular membranes with a diameter of 200 μm using different approaches to study the mechanical properties of atomically thin alumina substrate. Markutsya et al. [32] reported freely suspended layer-by-layer 25–70 nm thick polymer membranes by testing their micromechanical properties.

Figure 6.1 illustrates the closed-membrane type with the heater, electrode and the sensing layer with an insulating layer with either silicon dioxide or silicon nitride or both. Nitrided porous silicon is also a good material in building the membrane as it acts as a better dielectric [34]. Maccagnani et al. [34] use screen-printing technique to offer high mechanical strength. Si plug or a thin layer of silicon [35,36] underneath the heater area can also be used as an excellent thermal isolator with better temperature uniformity across the membrane. But the problem with this structure is that the device becomes delicate, and the etching processes are much more complicated as this can be obtained by electrochemical etch-stop technique. Si plug has been reported in several publications by Kloeck et al. [37] and also by Muller et al. [38]. Muller has discussed the formation of this layer at wafer level. Table 6.1 summarizes the thermal conductivity of different membrane materials.

6.3.2 Suspended-Membrane Hotplates

In the suspended-membrane type, spider-like support beams carry the hotplate with the sensing layer as shown in Figure 6.2. Under the hotplate, the substrate is etched off by producing a microcavity to provide a very good thermal isolation. In this case, etching is performed from the front side using standard etchant or by sacrificial etching [52]. The secondary advantage of this is that no photolithography is required from the backside. This is the reason for the popularity of this latter approach. Recently, a slightly modified approach has been reported by Dusco et al. [53]. Instead of etching the

TABLE 6.1
Thermal Conductivity of Different Membrane Materials

Materials	Specific Heat in J/kg-K	Density (kg/m³)	Thermal Conductivity at 300 K (W/m-k)	References
Silicon	700	2330	150	[39]
Silicon oxide	1000	2200	1.4	[40–42]
Silicon nitride	673	3290	30	[43–45]
Polyimide	702	2330	0.28	[46]
Porous silicon	850	466	1.2	[47]
Nitrided porous silicon	680	1552	0.74–4.09	[34]
Silicon oxynitride	97	2810	5	[48,49]
Silicon carbide	750	3100	120	[50]
Oxidized macroporous silicon	—	—	1	[51]

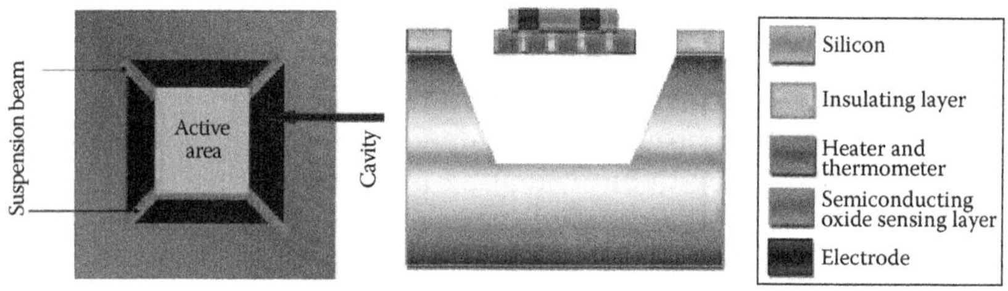

FIGURE 6.2
Schematic of a suspended-membrane-type gas sensor.

silicon substrate by a concentrated (some 30%) KOH solution, the silicon is made porous by HF in the first step and then the porous silicon is etched by a diluted (1%–2%) KOH solution. This device exhibits a high stability against mechanical shocks. The air can flow around the small hotplate and does not need to flow around the whole chip as in the case of the membrane-type solution. Another advantage of this structure is that this type of sensor exhibits much lower power consumption due to the exceptionally low thermal mass of the membrane. Though there are several positive aspects mentioned earlier, this type of membrane couldn't be proven well for mechanical instability as it is supported by only four beams. Belmonte et al. [54] made a trick to combine the suspended membrane fabricated on a backside bulk micromachined membrane.

A novel concept of fabricating microhotplate where the membrane and heaters both are made of differently doped silicon structures was reported recently [55]. The doped regions behave as resistive heaters, the geometry of which is selected to optimize the temperature distribution. Different designs using such microcantilever hotplate have been reported by Fujres et al. [56]

and Lee et al. [57]. Natalya et al. [58] presented cantilevers having a temperature tolerance of 2%–4% over the active area, which is a one-step approach for the betterment of the device till reported. Modelling as well as experimental verification has been done by Sebastian and Wiesmann [59].

6.4 Microheater Design Issues

A microheater is basically a small structure fabricated using a complementary metal oxide semiconductor (CMOS)-compatible process and should have the lowest possible power consumption. Microheaters comprise a hotplate on a micromachined silicon membrane.

The temperature rise is obtained by the Joule effect as in the case of normal resistor deposited on a membrane a few micrometres thick. The design of a microheater structure is very critical as the design would take into account the temperature distribution and power consumption issues. Different dimensions of the membrane and different geometries of the microheater structure have to be considered to find the best possible result. It is to be noted that the terms 'microheater' and 'microhotplate' are compatible. The key factors of a microhotplate design are as follows: (1) fast thermal response, (2) low power consumption, (3) temperature uniformity and (4) scalability. Depending on the requirement, factors have to be optimized. As, for example, fast response time is the main factor when the microheater is to be operated dynamically. Low power consumption is a very fundamental requirement particularly in field application where the sensor is battery operated. Uniform temperature distribution over the active area is another necessary requirement as it enhances the response of the sensor as the sensing area overlying the microhotplate is highly affected. Temperature uniformity minimizes the heater area that is very important for the miniaturization of the device and also heater reliability. The temperature uniformity depends on the membrane materials, membrane area and microheater geometry. The total power consumption can be minimized by the following:

1. Fabricating thin membranes of low thermal conductivity.
2. Adjusting the heater size (edge length 'a') to the size of the closed membrane (edge length 'b') as b/a. For best results, it should be 2:1.

The parameter that compares the side length of the membrane to the heater size (edge length) is known as the membrane-to-heater ratio (MHR). Poor MHR can lead to a process yield below 40% and higher power consumption. In the present work, MHR has been taken to be 2.

The stress generated by the thermal expansion of the stacked membrane due to heating of the sensor causes undesirable membrane deflection.

When heating due to the layer configuration of the microhotplate, the bimetallic effect produced by SiO_2 and silicon films deflects the structure. At a very high temperature, the thermally induced stress can be high enough to produce the structure crack. The maximum stress induced in the microheater is very high, which will obviously reduce the lifetime of the component. So in order to limit this value, the high stress over the different geometries of the microheater is investigated.

In a standard geometry of a microheater, due to electron accumulation in the concave corners of the heater coil, the localized temperature is increased, which may cause the generation of microcracks leading to shorter lifetime, whereas in the curved pattern, electron accumulation in the concave corners is reduced, which in turn lowers the localized temperature rise and also decreases the resistance of the heater coil, resulting in an increase of the maximum temperature rise.

Thus, the periphery of the active area seems to have a lower temperature than the inner part. This is the case that has been mostly observed with conventional meander-shaped heater.

6.5 Heater Material Selection

Choosing a perfect material for the microheater is a challenge for the better performance of the gas sensor. The right choice of an appropriate heating material is prevalent; it depends on so many factors like high electrical resistivity, low thermal expansion coefficient, low thermal conductivity, high Young's modulus, low cost, easy fabrication and, most importantly, compatibility to standard silicon fabrication technology. Various authors have reported mostly platinum and polysilicon as the heating element. The journey was started with Au [17] and Al [57] as heater elements. Gradually, it was found that these materials have so many limitations like low electrical resistivity making the length of the microheater longer, there is affinity towards oxidation even at room temperature, and there are poor contact formation and electromigration effect at a high temperature. The resistivity of polysilicon [60,61] is very high, making it suitable for a better heating element, but the problem is that it cannot be deposited by electron beam deposition technique. Rather, it has to be deposited using chemical vapour deposition technique, which is a quite expensive method. Platinum is a very popular high-temperature heating element, but the material itself is quite expensive, and electrical contact formation is also critical. Moreover, after 600°C, its resistivity changes [62–64]. Moreover, as Pt has a positive temperature coefficient, it produces amplifying effects on hot spots leading to a long-term reliability of the microheater structure. Researchers are continuously rendering their effort to find out some low-cost CMOS-compatible heating elements and ultimately to find out some alloys rather than the pure metal amalgamated with lots of positive features into a single material. Recently, nickel and nickel alloy–based microheater have also been reported [65,66]; they are now utilized for their lower TCE and low thermal conductivity. Alloys like NiCr [66], Dilver P1 [33], Mo [67] and tungsten [68] with some other semiconducting materials like doped silicon [58,69], silicon carbide [34], doped II–VI metal oxides like Sb-doped SnO_2 [70], titanium [71], titanium nitride [72] and hafnium diboride [73] have also been implemented as good heater elements. Some extra features like high Young's modulus, high yield, lower coefficient of thermal expansion, corrosion resistance, humidity resistance and non-magnetic property make them eligible to be used as a new microheater element. Molybdenum (Mo) has been reported as a microheater element due to its linear nature of electrical resistivity at very high temperature (~700°C) [67] than that of platinum. Ali et al. [68] reported novel high-temperature tungsten (W) resistive heaters on the silicon-on-insulator microhotplates. Tungsten has a tendency of oxide formation above 300°C, which is its main drawback. The use of a specific heater material is application specific depending on the requirement; therefore, there is no *thumb rule* for it. Instead, there is a compromise to make a good choice of a heater element.

TiN is very much suitable for high-temperature applications, but thermal stress may cause problem making it inappropriate for high temperature [72]. Table 6.2 gives the different properties of different heater materials.

6.6 Heater Geometry Selection

The crucial part of a semiconductor gas sensor is the microheater. Due to the adsorption/desorption kinetics, the resistance of the semiconducting metal oxides changes as a function of the varying concentration of the target gases. The area where this change is maximum is called the active area. To sense these resistive changes over the active area, the temperature of this region should attain a particular temperature range. It is worth mentioning that this active area is nothing but the area where the microheater lies. Hence, the sensitivity and response time of the sensor are highly related to the operating temperature of the microheater. So, their proper design is a crucial factor in this field. Moreover, a stable and uniform temperature over the active area implies a better performance of the sensing layer. Hence, temperature uniformity is a vital factor while designing the MEMS-based microheater.

Several researchers tried to avoid temperature non-uniformity over the active area by using a relatively thick silicon island or silicon plug beneath the single or composite (SiO_2/Si_3N_4) membrane. Some of them have tried to apply metalloid compounds on top of the heater, which conduct heat and electricity better than nonmetals but not like the metals. By immersing

TABLE 6.2

Comparison of Properties of Different Heater Materials

Materials	Electrical Resistivity in Ω-m	Thermal Expansion Coeff. α in 10^{-6}/K	Thermal Conductivity at 300 K in W/M-K	Melting Point (°C)	Density in kg/m³	Young's Modulus GPa	Poisson's Ratio	Specific Heat in J/kg-K	References
Gold	3.99×10^{-8}	14.1	180.93	1064	19,320	10.8	0.42	0.3	[17]
Al	2.7×10^{-8}	23.1	235	660	2,700	750	0.35	900	[57]
Polysilicon	32.2×10^{-8}	2.8	29–34	1412	2,330	169	0.22	753	[60,61]
Pt	10.6×10^{-8}	8.8	70	1772	2,109	168	0.38	130	[62–64]
Ni	6.8×10^{-8}	13	90.7	1453	8,900	200	0.31	440	[65]
NiCr	108×10^{-8}	14	15	1400	8,900	200	0.38	460	[66]
Ni alloy DilverP1	49×10^{-8}	4–5.2	17.5	1450	8,250	207	0.30	500	[33]
Molybdenum	5.5×10^{-8}	4.8	139	2623	2,623	329	0.31	250	[67]
Tungsten	5.5×10^{-8}	4.5	177	3410	19,300	411	0.284	140	[68]
Si/B	5	2.6	—	1414	2,330	—	—	750	[58,69]
Silicon carbide	31.6×10^{-8}	3.9	111	2730	3,210	390	0.15	667	[34]
Sb-doped SnO_2	6.5×10^{-3}	1	—	1919	—	—	—		[70]
Ti	42×10^{-8}	8.6	21.9	1668	4,506	116	0.32	455	[71]
TiN	81.3×10^{-8}	9.35	28.84	2930	5,430	250	—	—	[72]
Hafnium diboride	—	960	—	3250	11,100	—	—	—	[73]

the semiconductor substrate in a heated liquid for some time and then by baking, temperature uniformity may also be obtained. In achieving temperature uniformity over the active area, the simplest approach is the modification of heater geometry. Moreover, due to the heating effect, maximum stress induced in the microheater is very high, which will obviously reduce the lifetime of the component. So, in order to limit the high stress over the microheater, different geometries of microheater are investigated. There are several shapes that have been reported so far, e.g. (1) rectangular, (2) circular, (3) square and (4) asymmetrical shapes (like honeycomb or others) and (5) some 3D structures. Figure 6.3 gives the schematic of few microheater geometries though there are many other observations made by different researchers [74–76]. The most generic structure is the meander shape. Several new structures have been developed by modifying the meander shape, e.g. curved meander (where the convex corners have been

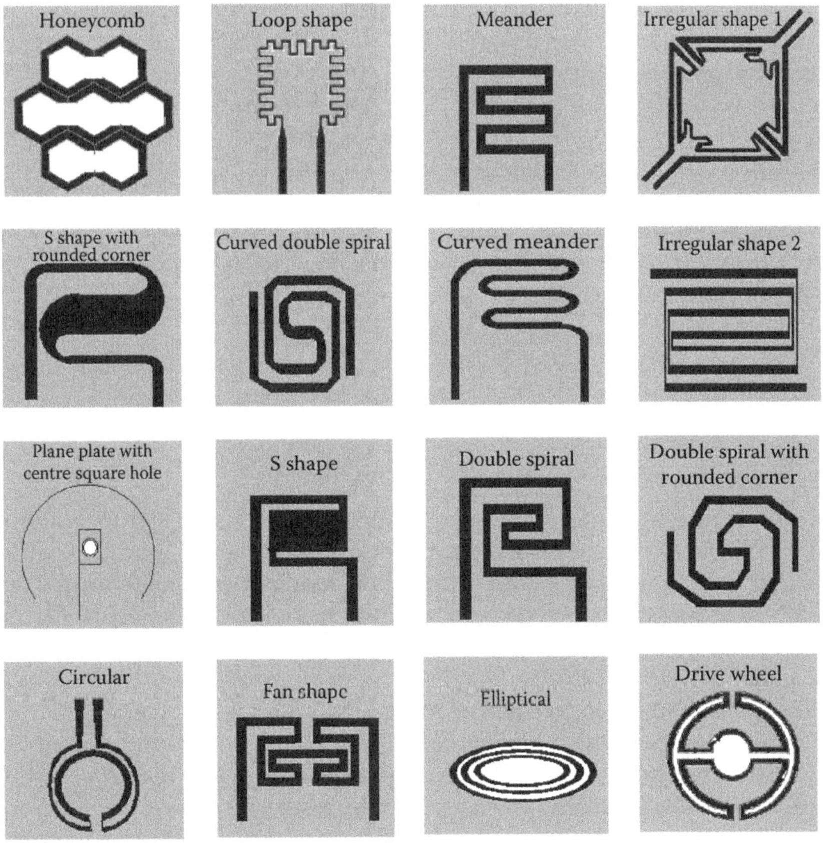

FIGURE 6.3
Different types of heater geometry.

rounded of) and S shape (where the number of turns is two). Different line widths and the spacing between the fingers have also resulted in some new structures. The blending of different patterns (e.g. square and rectangular) into a single heater has also resulted in a completely novel structure [57,77,78] and explored the area of innovative research. The simplest design ever reported is a square plate with a centre hole. Courbat et al. [79] fabricated a new structure, which is a combination of meander and double spiral, taking care of the reliability issue. Among all the structures, double spiral is preferred for its better temperature uniformity. But the sharp edges of these heater structure resulted in electron accumulation at the concave corners. This electron accumulation may cause rapid generation of microcracks, thus reducing the lifetime of the microheater. This can be overcome by rounding off the concave corners as reported by several researchers. Mele et al. [67] have reported the double spiral structure. Curved double spiral microheater with different line spacings has also been reported by Coppola et al. [80]. Ali et al. [68] have used tungsten-based circular-shaped microheater for its better mechanical stability using ZnO nanowire as the sensing material.

Graf et al. [76] have also reported that a circular-shaped microheater with long metal leads to improve the thermal isolation as compared to others. The circular design also supports the reduction of power consumption. Many researchers worked on the modified circular-shaped heater structure (e.g. simple half circular, drive wheel or elliptical design) [57,81]. Where the coplanar structures are used (heater and electrode sharing the same active area), often for better temperature uniformity and proper utilization of the active area, sometimes irregular-shaped microheaters are implemented [82]. But the designing of the mask gets complicated as it has no regular pattern.

6.7 Function of Interdigitated Electrode

The desirable features of a commercial gas sensor include significant sensitivity, specificity of detection, rapidity, less operator dependence, portability and low cost. It would be interesting if enrichment in sensitivity for such MEMS-based gas sensor can be done through some design modifications of the structure with ease of fabrication to detect as low as 100 ppm gas with considerably high sensitivity.

A conventional stacked structure integrated with microheater for gas sensing application generally employs catalytic noble metal electrode to enhance the sensitivity [6]. Currently, interdigitated electrodes (IDEs) are implemented in various sensing devices including surface acoustic wave (SAW) sensors, chemical sensors and MEMS GAS sensors. Comb-shaped contact is made with a motive to explore more probabilities of occupying a given sample space.

The output signal strength of IDEs is controlled through careful design of the active area, width and spacing of the electrode fingers. Here, the *active area* refers to the area where the field is associated, i.e. the area containing interdigitated parallel strips connected to the output meter. Min and Baeumner [83], for example, examined the relationship between signal-to-noise ratio and different aspects of IDE geometry and composition for optimizing oxidation and reduction reactions of potassium ferro-/ferrihexacyanide. It was found that increasing electrode thickness yielded an increase in current and overall signal with small increases in the signal noise [84]. This increase in current is thought to be due to increased surface area. Properly designed IDE must be provided in order to maximize the sensitivity of gas sensor measurements. The effect of the positioning of IDE on sensor platform is also an area of interest for many researchers [85,86].

It was very crucial to maintain the electrode temperature the same as the heater temperature for sensing gases. In the coplanar structure, this was better achieved by placing the IDE and microheater side by side sharing the same hotplate area. Moreover, this structure has the advantage of fabrication using single lithographic mask; thus, the number of fabrication steps is reduced.

Though it was found that increasing the number of electrode fingers and their finger widths yielded a proportional increase in overall signal [84], still if the spacing is too small, there are very high electric fields at the corners of the wide electrodes and the signal is less influenced. So, the largest square that can fit in the rectangular portion of the membrane maintaining the MHR equal to two has been considered to be the maximum area that can be covered by IDEs. This will exclude a narrow area near the periphery of the membrane region as non-uniformity may occur at the time of bulk micromachining.

6.8 Software Used

The numerical approximation of the temperature distribution, total heat loss, etc., can be obtained by replacing the sensor by a set of points. This set of points is called computational grid or mesh. The right choice of the computational grid is very important to obtain good approximation. Increasing the number of node points results in better accuracy but leads at the same time to an increase in computational time. Therefore, it is often better to use a non-uniform grid which models the parts of the sensor which are subjected to large changes with a narrow grid and regions with nearly no change in temperature with a rather rough grid.

A fundamental premise of using the finite element procedure is that the body is subdivided up into small discrete regions known as finite elements.

These elements are defined by nodes and interpolation functions. Governing equations are written for each element, and these elements are assembled into a global matrix. Loads and constraints are applied, and the solution is then determined. The finite element method (FEM) is a numerical tool to find out the approximate outcome of boundary value problems. It is mainly based on calculus of variations to produce a stable solution of the problem. The fundamentals of FEM are connecting many simple elemental equations over several small sub-domains, named finite elements, to approximate a more complex equation over a larger domain. In the year 1973, Strang and Fix extensively used mathematical formulae to implicate the FEM [87]. Later on, further emphasis was given to open-source finite element software programs. The method has since been generalized for the numerical modelling of physical systems in a wide variety of engineering disciplines, e.g. electromagnetism, heat transfer and fluid dynamics.

A typical workout of the method involves dividing the domain of the problem into a collection of sub-domains, with each sub-domain represented by a set of element equations to the original problem, followed by systematically recombining all sets of element equations into a global system of equations for the final calculation. The global system of equations has known solution techniques and can be calculated from the initial values of the original problem to obtain a numerical answer. In the first step, the element equations are simple equations that locally approximate the original complex equations to be studied, where the original equations are often partial differential equations.

The commercial finite element method (FEM) programs (FEM software) known as MEMS CAD tools that are generally used are Intellisuite, Comsol, Coventorware and ANSYS; these tools can be used in many engineering analysis, including thermal, structural, mechanical and electrical analyses. MEMS CAD tools are a solution for the design, simulation and optimization of MEMS devices. They optimize MEMS designs prior to fabrication, reducing prototype development cycle time and cutting manufacturing costs. Incorporating process templates, material data, mask layout and device analysis, CAD tools provide the platform for the entire design team to construct devices with higher yields. These softwares were mainly used to analyze the temperature uniformity, mechanical stability and thermal stability of the device.

6.8.1 Temperature Distribution

The heater geometry layout is the most pivotal role found in a gas sensor for the uniform temperature distribution over the membrane, though there are many other issues involved. Though meander-shaped microheater is mostly used for the experimental purpose, it offers non-uniform temperature distributions: the centre of the active area is at a higher temperature compared to the edges. There are many reports [39,40,75] indicating that an improved heater structure is necessary to maintain the temperature uniformity in the active region. However, optimizing the heater layout with FEM simulations is quite troublesome because it does not guarantee the optimum structure. An alternative solution as suggested by Gotz [39] and Graf [75] is the use of silicon plug of about 10 μm thickness just underneath the active area. It was proved that irrespective of the heater structure, a very uniform temperature distribution has been obtained due to the high thermal conductivity of silicon. Some researchers [40] used a relatively thin Si layer under the dielectric membrane producing a similar effect in addition to the increased robustness of the device with easy fabrication steps. Some authors fabricated a buried thin layer made of SiC [34] of high thermal conductivity or weakly doped diamond (between SiO_2-/Si_3N_4-stacked structures) as the heat spreader.

6.8.2 Mechanical Stability

Micromachined gas sensors are a challenge not only with respect to thermal design but also with respect to mechanical design. While doing the mechanical design, the following problems should be avoided:

- Large intrinsic or thermal-induced membrane stress leading to deformation or breaking of the membrane
- Plastic deformation at the metal–SnO_2/ZnO contact area
- Deformation/breaking of the membrane due to deposition of sensing layer
- Stress in the metallization sandwiched structure

To avoid these failures, the right set of process parameters has to be found, and the processes need to be well controlled. Very often, one will have to rely on trial and error, as thin-film mechanical properties depend strongly on microstructural characteristics like grain size, orientation, density and stoichiometry. Obviously, there are endless efforts to improve the mechanical behaviour of the MEMS platform with considerable temperature uniformity in the active region. One of the popular methods to minimize the residual stress is the use of SiO_2 and Si_3N_4 composite membranes, as the compressive stress developed in one is balanced by the tensile stress of the other. The technique used by other authors is to use the underlying thin silicon island/plug [39] or thin silicon membrane [40] to improve the mechanical stability of the hotplate.

Intrinsic stress in single layers or multilayer compositions can occur due to thermal stress and residual stress. Thermally induced stress is caused by mismatch of the thermal expansion coefficients of different films, which might lead to undesirable bimetallic wrapping effects, or is caused by non-uniform temperature distribution which might even vary with time. Puigcorbe et al. [74] reported that due to the different layer relationships of the stacked membrane, the bimetallic effect causes the structure to bend downwards.

However, the upward residual deflection of the membrane increases when applied voltages are increased and reached to a medium value. The latter points out that the thermal design and the mechanical design are strongly tied up with each other. Residual stresses can be led back to the fact that thin films as deposited are not in the most favourable energetic configurations resulting in compressive or tensile stress.

6.8.3 Thermal Response Time

Thermal transient response of a microheater can be obtained from a simple expression (Equation 6.1) by neglecting the actual temperature distribution inside the sensor and describing the thermal behaviour just with one overall thermal resistance R_{therm} and overall thermal capacity C. The balance between heating power P_{cl} acquired and lost through heat can be expressed by the following equation:

$$P_{cl} = \frac{C.dT(t)}{dt} + \frac{T(t) - T_{amb}}{R_{therm}} \tag{6.1}$$

where

$T(t)$ is the temperature of the microheater

T_{amb} is the ambient temperature

This equation allows the calculation of the temperature response to an arbitrary chosen heating power that acts as input; solutions can be found using Fourier and Laplace analyses. The temperature response to a step function is given by Simon et al. [22]:

$$T(t) - T_{amb} = P_{cl}.R_{therm}(1 - e^{-t/\tau}),\ \ \tau = R_{therm}.C \tag{6.2}$$

As seen from Equation 6.2, thermal time constant (τ) depends linearly on thermal resistance and capacity; obviously, the sensor with low thermal mass and small thermal resistance shows the fastest response. Always, the choice of thermal resistance will be a trade-off between power consumption and fast thermal response due to the fact that this leads to fast thermal response as well as an increase in power consumption. The thermal capacities of common sensor materials hardly differ as the minimization with respect to thermal capacity can only be achieved by the reduction of heated mass, i.e. minimization of the active area and membrane thickness. It can be concluded that the small suspended membranes have the fastest transient response than the closed membrane of the same dimension. It is noticed that the thermal time constant is indirectly temperature dependent. The use of low thermal capacity materials for membrane can improve the thermal time constant [88].

6.9 Heating Power Consumption

A microheater loses its temperature in three ways as depicted in Figure 6.4.

Radiation to the air: It is well known that radiation does not need any medium. Radiation is nothing but a heat transfer. This chapter particularly deals with thermal radiation, which is radiation emitted by matter with temperatures above zero Kelvin. Heat flows through radiation and occurs jointly with convection from the surface of the sensing layer to the surrounding gas. According to Yunus et al., 'radiation is a volumetric phenomenon, and all solids, liquids, and gases emit, absorb, or transmit radiation to varying degrees. However, radiation is usually considered to be a surface phenomenon for solids that are opaque to thermal radiation such as metals since the radiation emitted by the interior regions of such material can never reach the surface, and the radiation incident on such bodies is usually absorbed within a few microns from the surface' [89].

Heat loss by radiation is calculated by

$$\text{Heat flux, } q = \sigma\,(T_h^{\,4} - T_{amb}^{\,4}) \tag{6.3}$$

where

σ = Stefan–Boltzmann constant = 5.67×10^{-8} W/m^{2}K^4

$q = 5.67 \times 10^{-8}\,(426^4 - 300^4)$ W/m^2

$= 1.4136$ mW/μm^2

T_h = required temp.

T_{amb} = ambient temp.

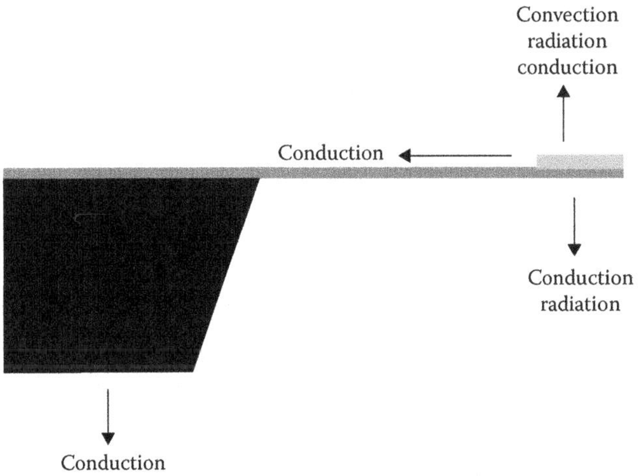

FIGURE 6.4
Heat evacuation system.

Natural convection: Heat transfer that occurs between a solid surface and the surrounding fluid is termed as convection. Heat flows from the surface of the sensing layer to the surrounding gas through convection. Convection is a mechanism that is done partly by conduction and partly by a mechanism of fluid motion [90]. There are two types of convection: one is forced convection and another is natural convection. In forced convection, the motion of the mass/fluid is caused by some external energy. On the other hand, the motion of the mass/fluid is caused by the gravitational force in natural convection. There exists a density gradient in a fluid due to temperature gradient. The gravitational force makes the flow from the more dense to the less dense medium, same as the flow of heat from a hot to cold surface until the temperature gradient becomes the same.

The heat transfer rate in natural convection is expressed by Newton's law of cooling as

$$q_{\text{conv}} = h \cdot A(T_s - T_{\text{amp}}) \tag{6.4}$$

taking a value of $h = 90$ W/m²K.

Then, the heat loss by convection is

$$q_{\text{conv}} = 90.(426 - 300) = 11.340 \text{ mW/}\mu\text{m}^2 \text{ (if area is unity)}$$

Conduction through the membrane: A 3D structure has to be considered for the calculation of heat conduction through the closed membrane, which is quite difficult to realize. Usually, the heat conduction perpendicular to the membrane is neglected, due to the small dimensions of the membrane.

In case of a suspended membrane, a 3D problem further reduced to a 1D problem as heat conduction occurs basically only in one direction, i.e. along the suspension beams with length L and cross-sectional area A_{beam}. For the suspended membrane, the heat loss through four suspension beams is

$$\begin{aligned} q_{\text{mem}} &= \frac{4\left(k \cdot A_{\text{beam}}\right)\left(T_{\text{hot}} - T_{\text{amb}}\right)}{L} \\ &= k \cdot G_{\text{m}}\left(T_{\text{hot}} - T_{\text{amb}}\right) \left[k \text{ is the thermal conductivity of the membrane}\right] \end{aligned} \tag{6.5}$$

where

$$G_{\text{m}} = \frac{4A_{\text{beam}}}{L} \tag{6.6}$$

For closed membranes, a simple model has been drawn for the ease of analysis where the rectangular membrane is substituted by a round one as

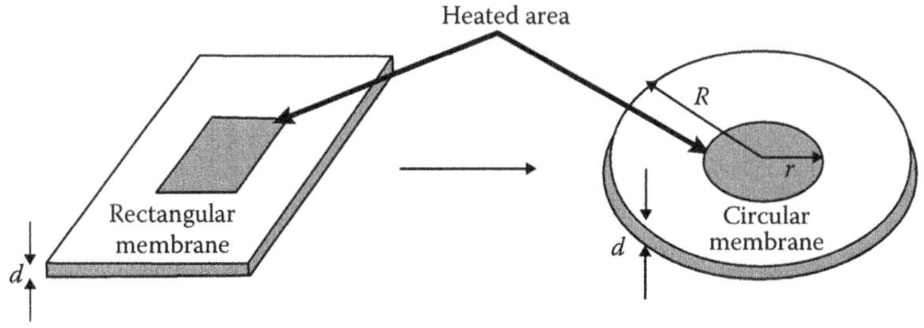

FIGURE 6.5
Approximation of 1D heat conduction through membrane.

depicted in Figure 6.5. This leads to a 1D heat conduction problem in cylindrical coordinates, which can be easily solved:

$$Q_{\text{membrane}} = \frac{2\pi kd(T_{\text{hot}} - T_{\text{amb}})}{\ln(R/r)}$$

$$= kG_{\text{m}}(T_{\text{hot}} - T_{\text{amb}}) \quad (6.7)$$

$$G_{\text{m}} = \frac{2\pi d}{\ln(R/r)} \quad (6.8)$$

where d denotes the thickness of the membrane and R and r are the radii of the membrane and the heated area, respectively.

It is clearly seen from Equations 6.5 and 6.7 that the choice of membrane/suspension beam of low thermal conductivity and small thickness is desirable to achieve low thermal losses.

The length-to-width ratio of the suspension beams or the relation R/r should additionally be chosen as large as possible to minimize heat losses. But whereas the suspension beams show a linear dependency on length-to-width ratio, the relation R/r influences the heat losses of a closed membrane in a logarithmic scale.

Convection through air may be ignored if the device dimension is too small and fluid (air) motion is also restricted within the packaged sensors with perforated encapsulation.

Radiation is dependent on temperature. Therefore, radiation losses can also be neglected if the heater is to operate in a low-temperature (300°C–400°C) regime. The matter has been taken into account first by Puigcorbe et al. [74]. Later on, however, due to T^4 dependency, they should be considered with care if very high sensor operation temperatures (~700°C–1000°C) are applied. As radiation linearly depends on material emissivity; often, low-emissivity materials (gold) are coated in the backside of the membrane to reduce radiation loss of high-temperature heaters [65].

6.10 Fabrication of Microheater

At the very beginning of the fabrication process, the wafer is first cleansed by the Radio Corporation of America (RCA) clean process. It is then followed by thinning of the Si wafer with CP4 solution. Oxidation is then performed followed by lithography at the backside of the wafer using the first mask. In lithography, positive photoresist is used. The oxide layer is then etched out to form the window for micromachining. Micromachining is performed by EDP solution, although it is very poisonous. The advantage of EDP is that it does not attack the oxide regions. After micromachining, Ni is deposited by E-beam evaporation. In the next step, again lithography is performed with the help of the second mask. Finally, selective etching of the Ni is done to form the microheater.

The fabrication of the microheater has been presented in this chapter. At the time of fabrication, a number of problems were encountered which have been solved with experience. In the lithography process, the proper alignment of the mask is very important. This can be solved by double sided lithography. During thermal oxidation, the rate of oxygen flow should be maintained at the proper rate. The rate of oxygen flow generally depends on the required thickness of the oxide layer and the volume of the furnace. During micromachining, the proper concentration of the etchant solution and proper temperature should be maintained. Since etching depends on time, the exact time should be allowed for micromachining of the sample. Highly clean environment is always a must for fabrication process.

Case Study

For proper understanding, the fabrication process that has been presented here is from one of the published papers [33] and is pictorially represented in Figure 6.6.

The fabrication starts by selecting a low-resistivity (10–20 cm) silicon <100> wafer (P-type). After RCA cleaning of the wafers, it is oxidized by thermal oxidation (generally at 1000°C), and a dry–wet–dry sequence for 30, 150 and 30 min, respectively, is done to obtain a 0.8 m thick insulation. The oxide is patterned in the backside of the wafer using positive photo resist (PPR) for subsequent micromachining. The window in the backside is chosen to obtain a 2 mm × 2 mm silicon membrane of 50 μ thickness. Micromachining using 8% TMAH at 80°C was then carried out from

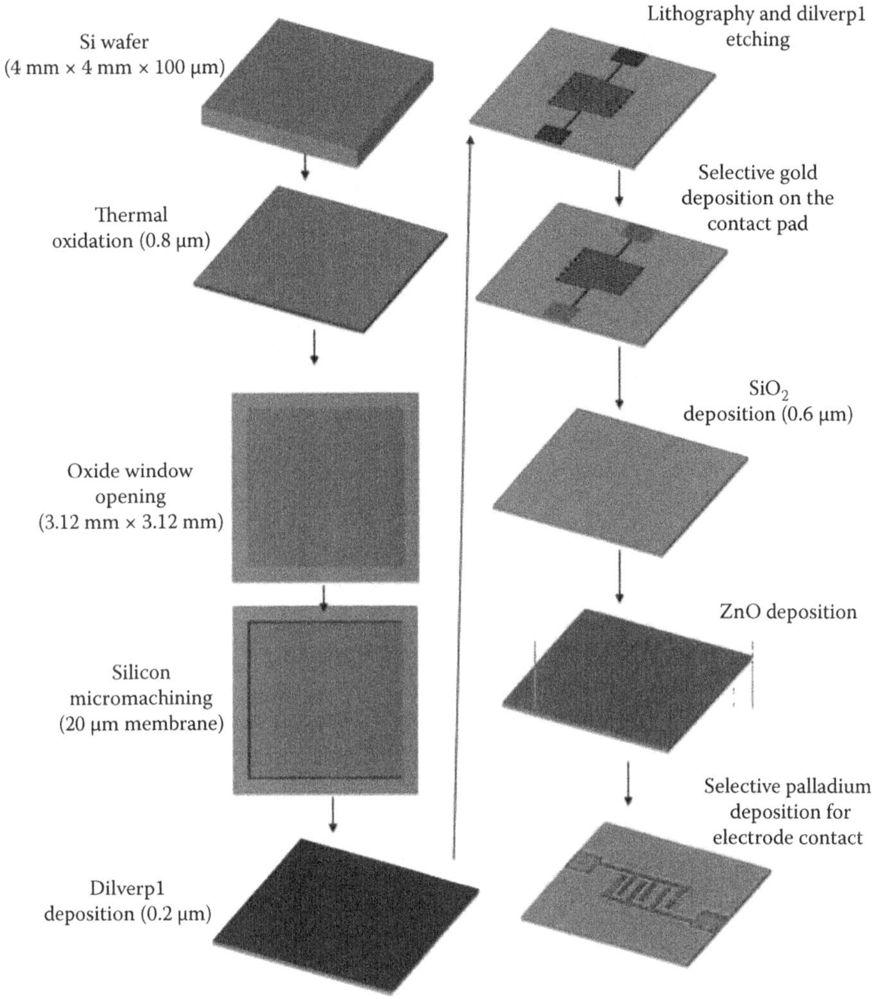

FIGURE 6.6
Pictorial representation of fabrication process.

backside in order to reduce the mass of the device, which will reduce the thermal loss of the device. Then, a 0.2 m thick Ni alloy is deposited (by e-beam deposition [5 kV, 55 mA]) at 10⁻ torr and annealed in N_2 ambient at 200°C for 45 min. The Ni alloy was then patterned using etch-back technique to obtain a meander-shaped constant pitch (100 m) microheater with a line width of 100 m. As our structure allows for a coplanar design, the IDE (line width 70 m and spacing also 70 m) is also defined in the same lithographic step. The etching is done using a solution of 10% HNO_3 acid buffered with acetic acid in a proportion of 1:3

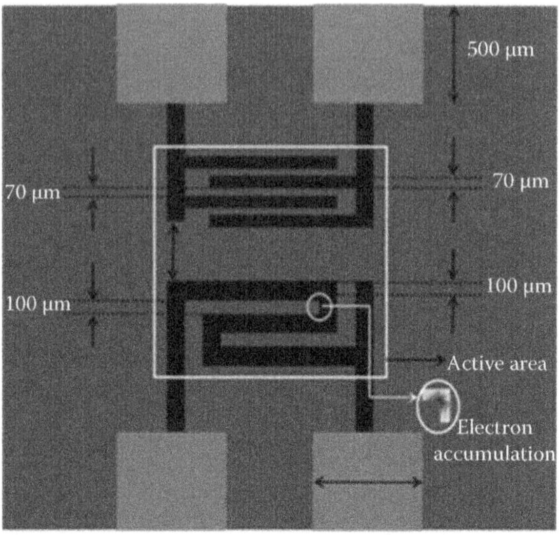

FIGURE 6.7
Schematic of the microheater structure. (Adapted from Roy, S. et al., *Sens. Lett.*, 9, 1382, 2011.)

for HNO_3/CH_3COOH. It is interesting to note here that the thickness of the NPR used in the lithographic step plays a crucial role in the etching process. The contact pads of the microheater and the IDE are then electroplated with gold for subsequent wire bonding. The gold electroplating is necessary for reliable gold wire bonding. Gold electroplating is carried out in a cyanide bath of $KAu(CN)_2$ at a current density of 0.3–0.5 mA/cm^2. The selective electroplating of the pads is obtained by performing another lithographic step to mask the microheater and the IDE. The FESEM image of the microheater is shown in Figure 6.7. From the fabricated microheater, the room temperature resistance was found to be 180 Ohm (approximately).

6.11 Microheater Array

It has been observed by almost all researchers working with oxide semiconductors for gas sensing that the operation of such sensors with selectivity for a particular gas is extremely difficult, especially when the changes in the electrical properties are used as the sensor signal. The use of sensor arrays and artificial neural network can normally solve this problem.

Microheater array is the generic structure of different numbers of heater combined, and it can be designed to detect the individual gases from a mixture of true gases by coating with different sensing films at a certain time with its different sensor block devices [91,92]. The sensing films are generally oxides or metal oxide compositions and allow the determination of information about the ambient gases. The required temperature for the gas sensor is generally ~150°C–300°C, as these temperatures provoke many factors related to the sensing film that are generally semiconducting metal oxides (ZnO, SnO_2, etc.) and play a significant role in determining the response to the target gases like CH_4, CO and NO_2 [93–95].

Microheaters generate an optimum operating temperature for different semiconducting metal oxides. So, microheater array is more suitable than a single microheater for gas sensor devices.

MEMS-based integrated gas sensors provide several advantages for applications such as array fabrication, small size and unique thermal manipulation capabilities and uniform heating throughout the active area. Sufficient selectivity of metal oxide sensors has been achieved even if several sensors are put together to an array. Instead of array, the use of several separate sensor elements can solve the purpose but with an increased power consumption. Even though the use of array is very promising with respect to sensor selectivity, one has to have in mind that the use of sensor arrays leads to an increased size of the sensor device. Figure 6.8 shows one of the published structures.

FIGURE 6.8
SEM image of the microheater array structure. (Adapted from Prasad, M. et al., *Sens. Transd. J.*, 103, 44, 2009.)

References

1. J. G. Lu, P. C. Chang and Z. Y. Fan, Quasi-one-dimensional metal oxide materials-synthesis, properties and applications, *Mater. Sci. Eng. R* 52, 49–91 (2006).
2. J. W. Orton and M. J. Powell, The Hall effect in polycrystalline and powdered semiconductors, *Rep. Prog. Phys.* 43, 1263–1307 (1980).
3. A. Rothschild and Y. Komem, On the relationship between the grain size and gas-sensitivity of chemo-resistive metal-oxide gas sensors with nanosized grains, *J. Electroceram.* 13, 697–701 (2004).
4. C. Baratto, G. Sberveglieri, A. Onischuk, B. Caruso and S. di Stasio, Low temperature selective NO_2 sensors by nano structured fibers of ZnO, *Sens. Actuators B* 100, 261–265 (2004).
5. F. Lu, Y. Liu, M. Dong and X. P. Wang, Nanosized tin oxide as the novel material with simultaneous detection towards CO, H_2 and CH_4, *Sens. Actuators B* 66, 225–227 (2000).
6. S. G. Ansari, P. Boroojerdian, S. R. Sainkar, R. N. Karekar, R. C. Aiyer and S. K. Kulkarni, Grain size effects on H_2 gas sensitivity of thick film resistor using SnO_2 nanoparticles, *Thin Solid Films* 295, 271–276 (1997).
7. E. Kanazawa, G. Sakai, K. Shimanoe, Y. Kanmura, Y. Teraoka, N. Miura and N. Yamazoe, Metal oxide semiconductor N_2O sensor for medical use, *Sens. Actuators B* 77, 72–77 (2001).
8. K. Wallgren and S. Sotiropoulos, A Nafion based co-planar electrode amperometric sensor for methanol determination in the gas phase, *J. Chem. Sci.* 121, 703–709 (2009).
9. T. J. Hsueh, C. L. Hsu, S. J. Chang and I. C. Chen, Laterally grown ZnO nanowire ethanol gas *sensors, Sens. Actuators B* 126, 473–477 (2007).
10. Z. Yang, Y. Huang, G. Chen, Z. Guo and S. Cheng, Ethanol gas sensor based on Al-doped ZnO nanomaterial with many gas diffusing channels, *Sens. Actuators B* 140, 549–556 (2009).
11. R. Bene, I. V. Perczel, F. Reti, F. A. Meyer, M. Fleisher and H. Meixner, Chemical reactions in the detection of acetone and NO by a CeO_2 thin film, *Sens. Actuators B* 71, 36–41 (2000).
12. F. Dimeo Jr., S. Semancik, R. Cavicchi, J. Suehle, N. Tea, M. Vaudin and J. Kelliher, Silicon microhotplate arrays as a platform for efficient gas sensing thin film research, *Mater. Res. Soc. Symp. Proc.* 444, 203–208 (1997).
13. J. Puigcorbe, D. Vogel, B. Michel, A. Vila, A. Gracia and C. Cane, Thermal and mechanical analysis of micromachined gas sensor, *J. Micromech. Microeng.* 13, 548–556 (2003).
14. C. Rossi, P. Temple-Boyer and D. Esteve, Realization and performance of thin SiO_2/SiNx membrane for microheater applications, *Sens. Actuators A* 64, 241–245 (1998).
15. C. Rossi, E. Scheid and D. Esteve, Theoretical and experimental study of silicon micromachined microheater with dielectric stacked membranes, *Sens. Actuators A* 3, 183–189 (1997).
16. J. W. Judy, Microelectromechanical systems (MEMS): Fabrication, design and applications, *Smart Mater. Struct.* 10, 1115–1134 (2001).

17. J. Puigcorbe, A. Vila, J. Cerda, A. Cirera, I. Gracia, C. Cane and J. R. Morante, Thermo-mechanical analysis of a microdrop coated gas sensor, *Sens. Actuators A* 97–98, 379–385 (2002).
18. A. Gotz, I. Gracia, C. Cane and E. Lora-Tamayo, Thermal and mechanical aspects for designing micromachined low power gas sensor, *J. Micromech. Microeng.* 7, 247–249 (1997).
19. R. Cavicchi, J. Suehle, K. Kreider, B. Shomaker, J. Small and M. Gaitan, Growth of SnO_2 films on micromachined hotplates, *Appl. Phys. Lett.* 66, 812–816 (1995).
20. J. O. Dennis, A. Yousif, and M. N. Mohamad, Design, Simulation and modeling of a micromachined high temperature microhotplate for application in trace gas detection, *Int. J. Eng. Technol.* 10, 67–74 (2010).
21. Y. K. Kato, R. C. Myers, A. C. Gossard, and D. D. Awschalom, Observation of the spin hall effect in semiconductors, *Science*, 306, 1910–1913 (2004).
22. I. Simon, N. Barsan, M. Bauer and U. Weimar, Micromachined metal oxide gas sensors: Opportunities to improve sensor performance, *Sens. Actuators B* 73, 1–26 (2001).
23. J. Gardner, A. Pike, N. de Rooji, M. Koudelka-Hep, P. Clerc, A. Hierlemann and W. GoÈpel, Integrated array sensor for detecting organic solvents, *Sens. Actuators B* 26/27, 135–139 (1995).
24. D. Lee, W. Chung, M. Choi and J. Back, Low-power micro gas sensor, *Sens. Actuators B* 33, 147–150 (1996).
25. G. Sberveglieri, W. Hellmich and G. MuÈller, Silicon hotplates for metal oxide gas sensor elements, *Microsyst. Technol.* 3, 183–190 (1997).
26. J. J. Vlassak and W. D. Nix, A new bulge test technique for the determination of the Young's modulus and the Poisson's ratio of the thin film, *J. Mater. Res.* 7, 3242–3249 (1992).
27. O. Tabata, K. Kawahata, S. Sugiyama and I. Igarishi, Mechanical property measurements of thin films using load-deflection of composite rectangular membranes, *Sens. Actuators* 20, 135–141 (1989).
28. H. Ni, H.-J. Lee and A. G. Ramirez, A robust two-step etching process for large-scale microfabricated SiO_2 and Si_3N_4 MEMS membranes, *Sens. Actuators A* 119, 553–558 (2005).
29. G. Benn, Design of a silicon carbide micro-hotplate geometry for high temperature chemical sensing, MS thesis, MIT, Cambridge, MA, 2001.
30. Y. Xiang, T. Y. Tsui, J. J. Vlassak and A. J. McKerrow, Measuring the elastic modulus and ultimate strength of low-k dielectric materials by means of the bulge test, *IEEE International Interconnect Technology Conference*, San Francisco CA, 2004.
31. M. K. Tripp, C. Stampfer, D. C. Miller, T. Helbling, C. F. Herrmann, C. Hierold, K. Gall, S. M. George and V. M. Bright, The mechanical properties of atomic layer deposited alumina for use in micro- and nano-electromechanical systems, *Sens. Actuators A* 130, 419 (2006).
32. S. Markutsya, C. Jiang, Y. Pikus and V. V. Tsukruk, Freely suspended layer by layer nanomembranes: Testing micromechanical properties, *Adv. Funct. Mater.* 15, 771 (2005).
33. S. Roy, T. Majhi, A. Kundu, C. K. Sarkar and H. Saha, Design, fabrication and simulation of coplanar microheater using nickel alloy for low temperature gas sensor application, *Sens. Lett.* 9, 1382–1389 (2011).

34. P. Maccagnani, L. Don and P. Negrini, Thermo-insulated microstructures based on thick porous silicon membranes, *Proceedings of the 13th European Conference on Solid-State Transducers,* The Hague, the Netherlands, 1999, pp. 817–820.
35. A. Gotz, I. Gracia, C. Cane, E. Lora Tamayo, M. C. Horrillo, J. Getino, C. Garcıa and J. Gutierrez, A micromachined solid state integrated gas sensor for the detection of aromatic hydrocarbons, *Sens. Actuators B* 44, 483–487 (1997).
36. S. Roy, C. K. R. Sarkar and P. Bhattacharyya, Ultrasensitive pd–ag/zno/nickel alloy-based metal–insulator–metal methane sensor on micromachined silicon substrate, *IEEE Sens. J.* 12, 2526–2527 (2012).
37. B. Kloeck, S. D. Collins, N. F. de Rooij and R. L. Smith, Study of electrochemical etch-stop for high precision thickness control of silicon membranes, *IEEE Trans. Electron. Dev.* 36(4), 663–666 (April 1989).
38. S. Muller, J. Lin and E. Obermeier, Material and design considerations for low power microheater modules for gas-sensor applications, *Sens. Actuators B* 24/25, 343–346 (1995).
39. L. Sheng, Z. Tang, J. Wu, P. C. H. Chan and J. K. O. Sin, A low-power CMOS compatible integrated gas sensor using maskless tin oxide sputtering, *Sens. Actuators B* 49, 81–87 (1998).
40. S. Roy, C. K. Sarkar and P. Bhattacharyya, Low temperature fabrication of a highly sensitive methane sensor with embedded Co-planar nickel alloy micro-heater on MEMS platform, *Sens. Lett.* 10, 759–768 (2012).
41. S. Roy, C. K. Sarkar and P. Bhattacharyya, A highly sensitive methane sensor with nickel alloy microheater on micromachined Si substrate, *Solid State Electron.* 76, 84–90 (2012).
42. C. C. Lu and K. H. Liao, Microfabrication and chemo resistive characteristics of SBA-15-templated mesoporous carbon gas sensors with CMOS compatibility, *Sens. Actuators B* 143, 500–507 (2010).
43. D. Barrettino, M. Graf, S. Taschini, S. Hafizovic, C. Hagleitner and A. Hierlemann, CMOS monolithic metal oxide gas sensor microsystems, *IEEE Sens. J.* 6, 276–286 (2006).
44. Y. W. Lai and J. E. Lee, In situ study of thermal deformation of metal resistive heater on silicon nitride membrane by digital holographic microscopy, *Seventh IEEE International Conference on Nano/Micro Engineered and Molecular Systems, NEMS,* Kyoto, Japan, 2012, pp. 557–561.
45. M. Blaschke, T. Tille, P. Robertson, S. Mair, U. Weimar and H. Ulmer, MEMS gas-sensor array for monitoring the perceived car-cabin air quality, *IEEE Sens. J.* 6, 1298–1308 (2006).
46. J.-S. Han, Z.-Y. Tan, K. Sato and M. Shikida, Thermal characterization of micro heater arrays on a polyimide film substrate for fingerprint sensing applications, *J. Micromech. Microeng.* 15, 282 (2005).
47. S. Mdler, J. Lin and E. Obermeier, Material and design considerations for low-power microheater modules for gas-sensor applications, *Sens. Actuators B* 24/25, 343–346 (1995).
48. S. Astie, A. M. Gue, E. Scheid and J. P. Guillemet, Design of a low power SnO_2 gas sensor integrated on silicon oxynitride membrane, *Sens. Actuators B* 67, 84–88 (2000).
49. A. Friedberger, P. Kreisl, E. Rose, G. Muller, G. Kuhner, J. Wollenstein and H. Bottner, Micromechanical fabrication of robust low-power metal oxide gas sensors, *Sens. Actuators B,* 93, 345–349 (2003).

50. W. Moritz, V. Fillipov, A. Vasiliev and A. Terentjev, Silicon carbide based semiconductor sensor for the detection of fluorocarbons, *Sens. Actuators B* 58, 486–490 (1999).
51. B. Mondal, P. K. Basu, P. Bhattacharya, C. Roychoudhury, B. T. Reddy and H. Saha, Oxidized macro porous silicon layer as an effective material for thermal insulation in thermal effect Distribution on a MEMS Micro-hotplate, *Third Asia Symposium on Quality Electronic Design*, San Jose, CA, 2011.
52. B. U. Moon, J. M. Lee, C. H. Shim, M. B. Lee, J. H. Lee, D. D. Lee and J. H. Lee, Silicon bridge type micro-gas sensor array, *Sens. Actuators B* 108, 271–277 (2001).
53. C. S. Ducso, E. Vazsonyi, M. Adam, I. Barsony and J. G. E. Gardeniers, A porous silicon bulk micromachining for thermally isolated membrane formation, *Eurosensors X, Proceedings*, Leuven, Belgium, 1995, pp. 227–230.
54. J. C. Belmonte, J. Puigcorbe, J. Arbiol, A. Vila, J. R. Morante, N. Sabate, I. Gracia and C. Cane, High-temperature low-power performing micromachined suspended micro-hotplate for gas sensing applications, *Sens. Actuators B* 114, 826–835 (2006).
55. T. Iwaki, J. A. Covington, F. Udrea, S. Z. Ali, P. K. Guha and J. W. Gardner, Design and simulation of resistive SOI CMOS micro-heaters for high temperature gas sensors, *J. Phys.* 15, 27–32 (2005).
56. P. Fujres, C. Ducso, M. Adam, J. Zettner and I. Barsony, Thermal characterization of micro-hotplates used in sensor structures, *Superlattice Microst.* 35, 455–464 (2004).
57. S. M. Lee, D. C. Dyer and J. W. Gardner, Design and optimization of a high-temperature silicon micro-hotplate for nanoporous palladium pellistors, *Microelectron. J.* 34, 115–126 (2003).
58. L. Natalya, W. Privorotskaya and W. P. King, Microsystem technologies: Micro- and nanosystems—Information storage and processing systems, *Microsyst. Technol.* 15, 333–340 (2009).
59. A. Sebastian and D. Wiesmann, Modeling and experimental identification of silicon microheater dynamics: A systems approach, *J. Microelectromech. Syst.* 17, 911–920 (2008).
60. J. Laconte, C. Dupont, D. Flandre and J. P. Raskin, SOI CMOS compatible low-power microheater optimization for the fabrication of smart gas sensors, *IEEE Sens. J.* 4, 670–680 (2004).
61. Z. Tang, Lie-Yi Sheng, C. H. Philip Chan, and J. K. O. Sin, A CMOS compatible integrated gas sensor, *Electron Devices Meeting*, IEEE Hong Kong, 9–12 (1996).
62. A. Pike and J. W. Gardner, Thermal modeling and characterization of micro-power chemoresistive silicon sensors, *Sens. Actuators B* 45, 19–26 (1997).
63. J. O. Dennis, A. Yousif and M. N. Mohamad, Design, simulation and modeling of a micromachined high temperature microhotplate for application in trace gas detection, *Int. J. Eng. Technol.* 10, 12–15 (2009).
64. G. Wiche, A. Berns, H. Steffes and E. Obermeier, Thermal analysis of silicon carbide based micro hotplates for metal oxide gas sensors, *Sens. Actuators A* 123/124, 12–17 (2005).
65. P. Bhattacharyya, P. K. Basu, B. Mondal and H. Saha, A low power MEMS gas sensor based on nanocrystalline ZnO thin films for sensing methane, *Microelectron. Reliab.* 48, 1772–1779 (2008).
66. J. Rolke, Adhesion measurements of thin films, *Electrocomp. Sci. Technol.* 9 51–57 (1981).

67. L. Mele, F. Santagata, E. Iervolino, M. Mihailovic, T. Rossi, A. T. Tran, H. Schellevis, J. F. Creemer and P. M. Sarro, A molybdenum MEMS microhotplate for high-temperature operation, *Sens. Actuators A* 148, 173–180 (2012).
68. S. Z. Ali, F. Udrea, W. I. Milne and J. W. Gardner, Tungsten-based SOI microhotplates for smart gas sensors, *J. Micromech. Syst.* 17, 1408–1417 (2008).
69. J. Lee and W. P. King, Microcantilever hotplates: Design, fabrication, and characterization, *Sens. Actuators A* 136, 291–298 (2001).
70. J. Spannhake, A. Helwig, G. Muller, G. Faglia, G. Sberveglieri, T. Doll, T. Wassner and M. Eickhoff, SnO_2:Sb—A new material for high temperature MEMS heater applications: Performance and limitations, *Sens. Actuators B* 124, 421–428 (2007).
71. G. S. Chung and J. M. Jeong, Fabrication of micro heaters on polycrystalline 3C-SiC suspended membranes for gas sensors and their characteristics, *Microelectron. Eng.* 87, 2348–2352 (2010).
72. J. F. Creemer, D. Briand, H. W. Zandbergen, W. V. Vlist, C. R. Boer, N. F. de Rooi and P. M. Sarro, Microhotplates with TiN heaters, *Sens. Actuators A* 148, 416–421 (2008).
73. F. Solzbacher, C. Imawan, H. Steffes, E. Obermeier and H. Moller, A modular system of SiC-based microhotplates for the application in metal oxide gas sensors, *Sens. Actuators*, 64, 95–101 (2000).
74. J. Puigcorbe, A. Vilà and J. R. Morante, Thermal fatigue modeling of micromachined gas sensors, *Sens. Actuators B* 95, 275–281 (2003).
75. M. Graf, D. Barrettino, M. Zimmermann, A. Hierlemann, H. Baltes, S. Hahn, N. Barsan and U. Weimar, CMOS monolithic metal-oxide sensor system comprising a microhotplate and associated circuitry, *IEEE Sens. J.* 4, 9–16 (2004).
76. M. Graf, R. Jurischka, D. Barrettino and A. Hierlemann, 3D nonlinear modeling of microhotplates in CMOS technology for use as metal-oxide-based gas sensors, *J. Micromech. Microeng.* 15, 190–200 (2005).
77. O. Sidek, M. Z. Ishak, M. A. Khalid, M. Z. Bakar and M. A. Miskam, Effect of heater geometry on the high temperature distribution on a MEMS microhotplate, *Third Asia Symposium on Quality Electronic Design*, San Jose, CA, 2011.
78. R. Phatthanakun, P. Deelda, W. Pummara, C. Sriphung, C. Pantong and N. Chomnawang, Design and fabrication of thin-film aluminum microheater and nickel temperature sensor, *Seventh IEEE International Conference on Nano/Micro Engineered and Molecular Systems, NEMS*, Kyoto, Japan, 2012, pp. 112–115.
79. J. Courbat, D. Briand and N. F. de Rooij, Reliability improvement of suspended platinum-based micro-heating elements, *Sens. Actuators A* 142, 284–291 (2008).
80. G. Coppola, V. Striano and P. Maccagnani, A nondestructive dynamic characterization of a microheater through digital holographic microscopy, *J. Microelectromech. Syst.* 16, 669–667 (2007).
81. M. Graf, D. Barrettino, H. P. Baltes and A. Hierlemann, *CMOS Hotplate Chemical Microsensors*, Eds. H. Fujita and D. Liepmann, Springer, New York, 2007.
82. F. Rastrello, P. Placidi and A. Scorzoni, A system for the dynamic control and thermal characterization of ultra low power gas sensors, *IEEE Trans. Instrum. Meas.* 60, 282–289 (2011).
83. Solzbacher, C. Imawan, H. Steffes, E. Obermeier, and M. Eickhoff. A new SiC/HfB_2 based low power gas sensor, *Sensors and Actuators*, B 77 111–115 (2001).
84. J. Min and A. J. Baeumner, Characterization and optimization of interdigitated ultramicroelectrode arrays as electrochemical biosensor transducers, *Electroanalysis* 16 724–729 (2004).

85. S. M. Radke and E. C. Alocilja, Design and fabrication of a microimpedance biosensor for bacterial detection, *IEEE Sens. J.* 4, 434–440 (2004).
86. M. M. EL Gowini and W. A. Moussa, A finite element model of a MEMS-based surface acoustic wave hydrogen sensor, *Sensors*, 10, 1232–1250 (2010).
87. K. Chatterjee et al., The effect of palladium incorporation on methane sensitivity of antimony doped tin oxide. *Mater. Chem. Phys.* 81, 33–38 (2003).
88. G. Strang and G. Fix, *An Analysis of the Finite Element Method*, Wellesley-Cambridge Press, Wellesley, MA, 1973.
89. B. L. Zink and F. Hellman, Specific heat and thermal conductivity of low-stress amorphous Si–N membranes, *Solid State Commun.* 129, 199–204 (2004).
90. Y. A. Cengel and R. H. Turner, *Fundamentals of Thermal-Fluid Sciences*, International edition, McGraw-Hill, Boston, MA, 2001.
91. M. N. Ozisik, A general approach, *Heat Transfer* 107, 13 (1985).
92. R. Wei, D. Pedone, A. Zurner, M. Doblinger and U. Rant, Fabrication of metallized nanopores in silicon nitride membranes for single-molecule sensing, *Small*, 6, 1406–1414, (2010).
93. V. K. Khanna, M. Prasad, V. K. Dwivedi, C. Shekhar, A. C. Pankaj and J. Basu, Design and electrothermal analysis of a polysilicon microheater on a suspended membrane for use gas sensing, *Indian J. Pure Appl. Phys.* 45, 332–335 (2007).
94. A. Datta Gupta and C. Roy Chaudhuri, Design and analysis of MEMS based microheater array on SOI wafer for low power gas sensor applications, *Int. J. Sci. Eng. Res.* 3, 1–8 (2012).
95. C. Baratto, G. Sberveglieri, A. Onischuk, B. Caruso, and S. di Stasio, Low temperature selective NO_2 sensors by nanostructured fibres of ZnO, *Sens. Actuators B* 100, 261–265 (2004).
96. B. J. Kim and J. S. Kim, Gas sensing characteristics of MEMS gas sensor arrays in binary mixed-gas system, *Materials Chemistry and Physics*, 138, 366–374 (2013).
97. M. Prasad, V. K. Khanna and R. Gopal, Design and development of polysilicon-based microhotplate for gas sensing application, *Sens. Transd. J.*, 103, 44–51 (2009).

Section II

Sensor Applications

7

Semiconductors as Gas Sensors

7.1 Introduction

Solid-state chemical and gas sensors are becoming increasingly necessary in today's global climate to detect malicious release of poisonous or toxic substances present in the atmosphere. Personal safety issues in both industry and residential environments demand highly sensitive detection of gases such as CO_x, NO_x and NH_3.

The advantages of an all-solid-state gas detection system are low power consumption, integration with existing circuitry and miniaturization. Microelectromechanical systems (MEMS) technology has allowed for the integration of the gas sensor, heating element and temperature sensor on a standard Si wafer with easy integration into standard complementary metal oxide semiconductor (CMOS) circuitry. MEMS implementation is not an easy task; it should be optimized first for nanostructured sensing mechanisms.

However, one of the disadvantages of traditional semiconducting metal oxide gas sensors has been low gas sensitivity due to the limited surface-to-volume ratio. In addition, many of the ceramic and thin-film gas sensors must be operated at temperatures exceeding 500°C in order to improve sensitivity.

Nanomaterials offer an exciting possibility for improving solid-state gas detection. Nanostructured sensing materials such as nanowires, nanotubes and quantum dots offer an inherently high surface area [1–10]. In fact, carbon nanotubes have one of the best available surface-to-volume ratios [4]. The increased surface area leads to high sensitivity and fast response and often allows for lower operating temperatures. In addition, different types of nanomaterials have different sensitivities. It is possible to apply different nanomaterials on the same MEMS platform to detect more complex gas mixtures.

In order to determine what type and structure of nanomaterials would be best suited for gas detectors, researchers have designed and fabricated gas detectors using single-walled carbon nanotubes, multiwalled carbon nanotubes, ZnO nanowires and nanoparticles and GaN nanorods [11,12]. Gas sensitivity, temperature response, initial response time and recovery times are important parameters of a gas sensor. Different sensor structures have been studied by many researchers [1–10], which are included in this chapter. Also the sensing mechanism and performance of some solid-state gas sensors are reviewed in this chapter, which is still a grey area and has not been fully understood.

The analysis of various parameters of metal oxides and the search of criteria, which could be used during material selection for solid-state gas sensor applications, are also the main objectives of this chapter.

7.2 Development of Semiconductor Sensors

Owing to its high surface/volume ratio, increased surface activity and strong adsorption to the target gas molecules, nanomaterials are grasping all areas of application. So the nanoparticle-built semiconducting films are expected to be of good gas sensing performances and hence have attracted much attention in the past decades. For instance, the metal oxide semiconducting (MOS) thin films consisting of nanoparticles, such as nanotubes, nanowires, nanorods and nanorings, have been extensively studied in their gas sensing properties to different gases. Generally, such nanostructured thin films display much better sensing performances than the corresponding bulk materials. These nanostructured porous thin films are usually produced on the electrode-equipped substrates with the use of a nanoprecursor. It is hard to control the nanostructure of the film, which results in non-uniformity in film thickness and poor reproducibility in film production. These ultimately result in inconsistency of sensing performances and hamper the long-term reliability of thin-film sensors. So there is a compromise while choosing a specific sensing layer. A single semiconducting nanowire gas sensor is excellent for its high sensing performance; still the use is restricted due to its complicated device construction, high cost and non-repeatability. Many different facile synthesis approaches with low cost have already been implemented for the fabrication of nanostructure-based thin-film gas sensors with high performances [13–16].

While working with semiconductor p–n junctions, researchers discovered that the junction parameters were changing due to environmental gases. At that time, this change was completely unwanted and was creating a problem. Encapsulation solved the problem as it was no longer exposed to the outside environment. At a later stage, this problem clicked the mind of researchers and was made to utilize the sensitivity of the semiconductor junction as a gas detection device.

After 1968, N. Taguchi publicized a solid-state sensor for the detection of combustible hydrocarbon gases. In 1972, the International Sensor Technology (IST) in Irvine, California, first proposed a solid-state sensor for the detection of hydrogen sulphide in a range of 0–10 ppm. Few years later, different hazardous gas detection was made possible by IST researchers at low ppm levels.

Properly manufactured solid-state sensors offer a very long life expectancy. It is not abnormal to find fully functional sensors that were installed 30 years ago. Because catalytic sensors burn the gas being detected, the sensor material is consumed and changed in the process, and the sensor eventually burns out.

With solid-state sensors, on the other hand, gas simply *adsorbs* onto the sensor surface, changing the resistance of the sensor material. When the gas disappears, the sensor returns to its original condition. No sensor material is consumed in the process, and hence the solid-state sensors offer a long life expectancy.

7.2.1 Fundamentals of Semiconductor Sensors

In the presence of gas, the metal oxide causes the gas to dissociate into charged ions or complexes, which results in the transfer of electrons. The built-in heater, which heats the metal oxide material to an operational temperature range that is optimal for the gas to be detected, is regulated and controlled by a specific circuit.

A pair of biased electrode was embedded into the metal oxide to measure its conductivity change. The change in the conductivity of the sensor resulting from the interaction with the gas molecules is measured as a signal. Typically, a solid-state sensor produces a very strong signal, especially at high gas concentrations. There are different ways of making solid-state sensors, each arrangement making the sensor's performance characteristics different.

7.2.2 Classification of Semiconductor Sensors

It is very tricky to classify sensors as they work on the basis of different principles. They may be based on the transduction principle (resistive/conductometric, capacitive, etc.), measure (pressure, temperature, stress, etc.) and materials (semiconductors, PZT, oxides, etc.); based on technology (thin/thick film, MEMS) and applications (aerospace, industry, automobile); or simply based on the properties of the sensor (piezoelectric, magnetic, optical, etc.). As per our area of interest, the classification based on the working principle has been given as follows:

Working Principle	Type of Sensors	Variable Parameter	Refs.
Concentration of free charge carrier changes in conducting material	Conducting polymer sensor Conductance sensor Semiconductor gas sensor	Δv, Δi, $\Delta\sigma$	[17–19]
Polarization of fixed charge changes in insulating material	Capacitive sensor	ΔC	[20]
Control of the charge distribution at the insulator–semiconductor interface	MOSFET, FET sensor	Δi, Δv, $\Delta\sigma$	[21]
Change of the resonance frequency of a quartz resonator in the presence of volatile molecules	Mass-sensitive sensors Quartz microbalance (QMB) Surface acoustic wave (SAW)	Δf, Δm	[22]
Change of electrical property of the sensor due to heat generated by the combustion of inflammable gases	Thermal conductivity sensors Calorimetric sensor Seebeck effect sensors Pyroelectric sensors Pellistors or catalytic sensors	Δv, Δi, ΔT, Δq	[23]
Any one of the optical property changes (phase, intensity, wavelength, polarization) in the optical fibre by the presence of gas	Fibre-optic sensor	Δv, Δi, $\Delta\Phi$, Δn	[24]
Modification of the electrochemical potential induced by charge swapping in the oxidation or reduction reactions at the electrodes	Electrochemical sensor	Δv, Δi	[25]

7.2.3 Different Structures of Semiconductor Gas Sensors

7.2.3.1 Resistive-Type Metal Oxide–Based Gas Sensors

In this sensor, resistance is measured between the two contact electrodes, taken from the top of the sensing (metal oxide) layer that has been deposited on a non-conducting substrate like glass, alumina or SiO_2. The operating temperature of the sensor signifies the major variation of the sensor signal originating from the changes in the electronic conductivity which is caused by the charge transfer in the course of chemisorptions and catalytic reactions at the surface and at grain boundaries. The advantages of this sort of sensors are that it is easy to fabricate and is capable of direct measurement.

7.2.3.2 Schottky-Type Metal Oxide–Based Gas Sensor

The key issue of this type of sensor is that the functional characteristics of sensors, e.g. response magnitude and response time, have been improved by adopting Schottky structures with catalytic metal contact as electrode. Recently, Salehi et al. [24] reported a selective CO and a NO sensor based on

Au/porous GaAs Schottky junction. Schottky junction is formed due to the work function difference of two materials. In the sensor structure, this has been formed by depositing a metal of higher work function on and above the semiconducting metal oxide layer. Most of the catalytic noble metals (Pd, Pt, Rh, etc.) make Schottky junctions with the semiconducting metal oxides by depositing the contact electrode with the noble metal electrode, thus initiating a catalytic effect as well as the collection of carriers. Due to natural non-stoichiometry, the adsorbed atmospheric oxygen in the form of molecules or ions produces a dangling bond on the metal oxide surface. The hydrogen containing molecules present in the reducing gases like H2 and VOC's; undergo chemisorptions which cause disintegration of molecules on catalytic metal electrode as well as on the semiconductor surface, to produce atomic hydrogen. Chemisorption is based on electron transfer between adsorbent and adsorbate. The atomic hydrogen gradually diffuses into the metal/metal oxide junction and lowers down the catalytic metal work function which causes band bending, thus changing the surface resistance of the sensing material. This change can be measured also by I-V, C-V or other electrical modes.

An example of a Schottky junction device is the metal–insulator–metal (MIM) structure where one dielectric layer is sandwiched between two metal layers [25]. The metal layer may be of the same material or different material. In general, an oxide sensor with catalytic metal contact/metal oxide/ohmic contact configuration is more desired, because most of the electrons generated in the sensing catalytic electrode (catalytic metal, e.g. Pd, Pt) due to the gas interaction process can be collected by the second ohmic electrode without carrier recombination in the transport path. This is obviously due to the thin semiconducting oxide barrier and highly conducting ohmic interface. The double potential barriers are thus formed by two metallic layers due to the difference in work function on either side of the metal oxide layer. The advantage of this device over the planar structure is that the higher barrier height prevents the flow of free carriers when no potential is applied. A higher potential barrier contributes to high electrical resistance, which falls suddenly in the presence of toxic gases. This change in resistance in the presence or absence of gas is high enough to have a higher response magnitude. The sudden fall of resistance occurs due to the fast vertical transport of free carriers through the thin (1 μm) metal–semiconductor junctions as the two metal layers are stacked upon each other.

This phenomenon is not observed in a normal planar structure. That's why a MIM structure has a fast response time. Consequently, the recovery time is also short.

The MIM configuration was first dictated by Fonash et al. [10]. Though the advantage of this structure, using ZnO as the sandwich layer and Pd and Zn as the electrode metals, was reported by Basu et al. [9] as a room temperature

H_2 sensor, a fast-response H_2 sensor at an elevated temperature was reported by Hazra and Basu [26] using a $Pd/TiO_x/Ti\text{-}Al$ MIM structure.

7.2.3.3 Metal Oxide Homojunction Gas Sensor

Generally metal oxide semiconductors are n-type semiconductors, which is completely unintentional. Though there are exceptional cases, some metal oxides found in nature are p type, e.g. CuO and NiO. ZnO has also offered p-type conductivity [27] somewhere. Thus, the invention of the semiconducting metal oxide p–n homojunction has been proved with ZnO [28]. Hazra et al. [26] showed that these kinds of p–n ZnO homojunctions are sensitive to H_2. There is a significant shift in the forward bias current of the p–n junction in the reducing gas atmosphere.

7.2.3.4 Metal Oxide Heterojunction Gas Sensor

A heterojunction is made of two dissimilar metal oxides with different band gaps, which has explored a new kind of gas sensor structure. ZnO/CuO (ZnO is an n-type semiconductor and CuO is a p-type semiconductor) is a relatively widely explored heterojunction sensing material used for CO gas sensor. Hu et al. [29] showed that this material has an excellent response towards H_2S and alcohol.

It was found that a heterojunction gas sensor is beneficial over a homojunction sensor as the recombination effect is less than that of the homojunction due to dissimilar band gap energy. Efficiency is improved in this way.

The heterojunction formed with the n-type ZnO and p-type composite based on a mixture of $BaTiO_3/CuO/La_2O_3$ is used to detect NO_2 and CO_2 [30]. This heterostructure has increased resistance in NO_2 and small decrease in resistance in CO_2. Silicon-based heterojunctions like metal oxides/Si have been studied extensively [31]. Ling and Leach [32] recently reported on SnO_2/WO_3-heterojunction-based NO_2 sensors.

7.2.3.5 Mixed Metal Oxide Gas Sensors

Mixed oxides are recently explored material and showed their competency for the easy detection of gases. The basic difference with heterojunction is that mixed oxides lead to modification of the electronic structure of the system. The modification is in the area of the surface as well as bulk properties, bulk electronic structure, band gap, Fermi level position, transport properties, etc. The new surface property was observed due to the new grain boundaries of different chemical compositions. Mixed oxide systems can be classified into three categories. The first category comprises of distinct chemical compounds, e.g. ZnO–SnO [33]. The second category is those mixed oxides that form solid solutions. The TiO_2–SnO_2 is an example of a solid solution [34].

The third category includes those which are neither compounds nor solid solution, e.g. TiO_2–WO_3 [35].

7.2.3.6 MEMS Gas Sensors

Taguchi-type sensor was the first produced commercially available solid-state sensor with large power consumption (1 W). Later, screen-printed ceramic gas sensors have power consumption (500 mW) still not appropriate for commercial use due to mounting difficulty and selectivity. Power consumption being very high, it was not suitable for use in battery-operated instruments. An important technology called CMOS–MEMS had come into existence about 20 years ago mainly for less power consumption with greater sensitivity. Since then, it has been developed greatly for realizing various types of physical sensors and actuators. This idea was first introduced in 1993 by Suehle et al. [36] where he reported a MEMS gas sensor design using microhotplate based on CMOS processes (CMOS-μHP), which was later patented by Semancik et al. [37].

The sensitive layer of micromachined metal oxide gas sensors is deposited onto a thin dielectric membrane of low thermal conductivity, which provides good thermal isolation between the substrate and the gas-sensitive heated area on the membrane. Micro-platforms are providing various benefits to gas sensors. First, the power consumption can be kept very low due to reduced thermal mass of the micromachined substrate (typical values obtained lie in the range between 30 and 150 mW [43–45]). Another advantage is that the active area, which is much smaller than the bulk counterparts, is confined to the heater region; hence the substrate itself stays almost at ambient temperature. Due to its small size, the mounting of the sensor element becomes much easier and a signal conditioning unit can be integrated on the same substrate if desired. Another advantage is that microheaters and electrodes can be fabricated by a standard lithographic step applying a coplanar structure as reported by Saha et al. The finger width of the interdigitated electrode lying in the μm range has been achieved [38]. The gas-sensitive area can in this way be tremendously reduced. Sensor arrays which are often needed to overcome the poor selectivity of single sensor elements can be easily implemented by MEMS technology. Beyond that, the reduced thermal mass of each micromachined element allows rapid thermal programming which can be used to study the kinetics of the surface mechanism [39].

Moreover, power consumption can be reduced drastically, especially through a pulse-mode heating operation with a small duty ratio. By saving power consumption, sensor devices can be made drivable with a battery, which is beneficial to cordless or portable gas sensors. Third, the excellent heating and cooling characteristics can provide gas sensors with new functions. The thermal characteristics of micromachined gas sensors have to be optimized with respect to low power consumption, well-controlled temperature distribution over the sensing layer and fast transient response, if the sensor is intended to be temperature modulated. Recently, Puigcorbe et al. [40] published a report on thermal and mechanical (fatigue) analysis of micromachined gas sensors.

Very recently with an aim to integrate a nanodimensional sensing film to MEMS structure, Gong et al. [41] and Kim et al. [42] reported on micromachined nanocrystalline SnO_2-based chemical gas sensors, where detection of H_2 is possible with very fast response and recovery time.

7.3 What Is a Nanosensor?

Nanosize is the dimension of any particle in nanometre, i.e. 10^{-9} m – 1 nm = cluster of 10 hydrogen atoms or 5 silicon atoms. As the particle size is drastically reduced from bulk size, certainly there will be some changes in material properties. These features are illustrated as follows:

1. *Very high surface-to-volume ratio*: It is very well known that nanocrystalline materials offer high surface-to-volume ratio than bulk material. With particle size of about 30 nm, only 5% of the atoms are on the surface; as we reduce the particle size to 10 nm, we get 30% atoms, and as we go to even smaller dimensions (~3 nm), we achieve 50% of atoms on the surface. When the crystal size is decreasing, more and more surface is exposed. So the fraction of atoms in the grain boundary increases, and the grain boundaries contain a high density of defects like vacancies and dangling bonds, which can play an important role in the transport properties of electrons of nanomaterials. We get a large number of dangling bonds or unsatisfying bonds on the surface of the material as a result of which the coordination number becomes different from that of the bulk portion. In this way, we can increase the effective sensitive area and hence greatly enhance sensitivity. It has also been seen that there exist certain optimum values of particle size for certain target gas sensing material combination.
2. *Lowered operating temperature*: Since nanomaterials have higher numbers of adsorption sites due to dangling bonds, the operating temperature of the sensor is low. It has been experimentally seen that for ordinary thick-film microcrystalline sensors, the operating temperature may be as high as 350°C, but for nanocrystalline materials, it can be lowered down to 200°C.
3. *Precise control over sensitivity by size reduction*: By adjusting particle size, porosity and thickness of the film according to our needs, we can increase the sensitivity, selectivity and response time of our sensor.

7.3.1 Thin Film Sensors

Thin film can be defined as a layer of material which is only a few micrometres thick (<1 μm). It is not self-sustaining and needs a substrate to adhere. Thin-film technology required a much smoother surface finish of substrate (Si is preferred). Atoms/molecules are highly energized with heat or electrical power or by some other means and directly transported from the source to the substrate to make an atomically thin layer. In recent years, researchers are strongly investigating the gas sensing properties of nanostructured materials because of their high surface-to-volume ratio and exceptional reduction in the activation energy for chemisorption. These nanostructured materials when agglomerated form a micrometre-level thick sensing layer. Due to the high surface-to-volume ratio of this thin film, the gas sensing properties of nanostructures are markedly better compared to their conventional thick-film counterparts, and they have the advantages of greater sensitivity, faster response and lower operating temperatures (which is essentially useful in lowering the power requirements for possible integration with CMOS technology). Noble metal sensitization/modification has been a well-observed phenomenon for metal oxide sensors to improve the sensing parameters. In particular, towards explosive gases like methane or H_2 sensing, due to the catalytic effect of noble metals, the optimum operating temperature has been reduced. Several publications have focused on the catalytic property of Pd/Pd–Ag contacts to ZnO thin films [43–46] for the sensing of reducing gases. However, other papers [47] are more cantered around dispersing Pd metal nanoclusters on the thin films leading to electronic sensitization of the metal oxide sensing layer.

To make the thin film, 1D nanostructures recently have caught the eyes due to their great potential for fundamental studies of the roles of dimensionality and size in their physical properties as well as for their versatility of applications in the field of optoelectronic nanodevices and sensors.

The substrate characteristics play an important role in the morphology of thin film. Metal oxide gas sensors are thin-film sensors that are used in gas leak detection (CH4, H_2, etc.) and ambient air quality monitoring in traffic (CO, NO_x). The idea of using semiconductors as gas-sensitive devices leads back to1952 when Brattain and Bardeen first reported the gas-sensitive effect on germanium [48]. Later Seiyama found the gas sensing effect of metal oxides [49].

7.3.2 Thick-Film Sensors

A layer of material that is thicker than the micron depth layer is termed as thick film (5–20 μm). *Thick-film* technology does not use lithographic technique; rather it uses conductive, resistive and insulating pastes, deposited in patterns defined by screen printing onto a ceramic substrate at high temperature.

7.3.2.1 Thick-Film Materials

- Generally alumina is used as a substrate material.
- In this technology, different components are produced on the substrate by applying *paste* to produce the required conductor patterns and resistor values. This technology is called screen printing. *Screens* are made of stainless steel or polyester mesh, with a reasonably open weave to pass through. Pastes are generally palladium- and platinum-alloyed copper and nickel with high dielectric strength (10^7 V/m). Metal particles in the paste are bound together and with the substrate also.
- It provides good insulation resistance (10^{22} Ω m^{-2}).

Taguchi first invented semiconductor sensors based on metal oxides. Nowadays, most of the commercially available sensors are manufactured in screen-printing technique on small and thin ceramic substrates. Screen-printing technique has the advantage that thick films of metal oxide semiconductor sensors are deposited in batch processing thus lowering the production cost. The power consumption of screen-printed devices is in the range of 1–5 W, which is a major disadvantage. The mounting is also difficult. Screen printing technique has the advantage that thick films of metal oxide semiconductor sensors are deposited in batch processing thus lowering the production cost [50].

7.4 Solid-State Chemical Sensors

A chemical sensor is a device that transforms chemical information, ranging from the concentration of a specific sample component to total composition analysis, into an analytically useful signal. The chemical information, mentioned earlier, may originate from a chemical reaction of the analyte or from a physical property of the system investigated. On the contrary, a physical sensor is one where no chemical reaction takes place. Typical examples are those based upon the measurement of absorbance, refractive index, conductivity, temperature or mass change. Sometimes it may become difficult to judge whether a sensor is operating in physical mode or chemical mode. This is, for example the case when the signal is due to an adsorption process.

The chemical sensor mainly depends on two functions: (1) receptor function, which is concerned with the reaction of target gas molecules with the semiconducting metal oxide surface or how each crystal responds to the stimulant gas in problem, and (2) transducer function, where the semiconductor particles are gelatinous together in the sensor element [51]. The microstructure of these aggregates is considered to be very important for the transducer function. Each particle is connected with its neighbours either by grain boundary contacts or by necks. In the case of grain boundary contacts, electrons should move across the surface potential barrier at each boundary. The change of the barrier height makes the electrical resistance of the element dependent on the gaseous atmosphere. The resistance and the gas sensitivity are not dependent on the particle size. In necks, electron transfer between particles takes place through a channel which is formed inside the space charge layer at each neck. The width of the channel is dependent on the size of the neck and the depletion layer length (*L*), and its change with gases gives rise to the gas-dependent resistance of the element.

7.4.1 Metal Oxide Semiconductors

Gas sensors are of prime importance when it is exposed in potentially hazardous environment (like mines), military application, biomedical area and many others. A semiconducting gas sensor is effectively a gas-sensitive resistor. A sensing element is normally a semiconducting material having high surface-to-bulk ratio deployed on a heated insulating substrate between two metallic electrodes.

Reactions involving gas molecules can take place at the semiconductor surface to change the density of charge carriers available, i.e. the gas–solid interactions leading to the physical adsorption and chemisorptions modify the electron/hole density in a relatively shallow region near the surface. In some semiconducting metal oxides like titanium dioxide, bulk diffusion of defects determines the sensor response. As a result, the conductance of the device changes progressively with changing atmospheric composition. The sensing material is not an elemental semiconductor, rather semiconducting metal oxides having wide–band gap energy treated as semiconductors. These semiconducting oxides are the fundamentals of smart devices as both the structure and morphology of these materials can be controlled precisely, and so they are referred to as functional oxides. They have mainly two structural characteristics: cations with mixed valence states and anions with deficiencies. By varying either one or both of these characteristics, the electrical, optical, magnetic and chemical properties can be tuned giving the possibility of fabricating smart devices. The structures of functional oxides are very diverse and varied, and there are endless new phenomena and application. Such unique characteristics make oxides one of the most diverse classes of materials with properties covering almost all aspects of materials science and areas of physics such as semiconductors, superconductivity, ferroelectricity and magnetism. The effectiveness of semiconductor oxide–based gas sensors depends on several factors:

1. The receptor function, transducer function and utility. Receptor function concerns the ability of the oxide surface to interact with the target gas. Chemical properties of the surface oxygen of the oxides are responsible for this reaction in an oxide-based device [52].

2. The optimum operating temperature increases the molecular activation energy so sensitivity will increase. But temperature above the optimum value increases the phonon scattering or lattice vibration resulting in poor performance.
3. The catalytic properties of the surface, i.e. if the surface is modified by a noble metal (Pd, Pt, etc.) or acidic or basic oxides, then the transducer function will increase [52].
4. The electronic properties of the bulk oxide and the microstructure.

7.4.2 Nanocrystalline Metal Oxide Semiconductors

Nanocrystalline materials are single- or multiphase polycrystalline solids with a grain size of a few nanometres (10^{-9} m), typically less than 100 nm. Since the grain sizes are so small, a significant volume of the microstructure in nanocrystalline materials is composed of interfaces, mainly grain boundaries, i.e. a large volume fraction of the atoms resides in grain boundaries. Consequently, nanocrystalline materials exhibit properties that are significantly different from their conventional coarse-grained polycrystalline counterparts. At present, the very broad field of nanostructured materials includes (1) nanoparticles, (2) nanocrystalline materials and (3) nanodevices. The potential applications for various kinds of nanoscale materials include dispersions and coatings, high surface area materials and functional nanostructures. Metal oxide semiconductors containing the grain size as small as 1 nm are called nanocrystalline metal oxide semiconductors [53]. The high surface-to-volume ratio makes it useful for different applications, e.g. optoelectronic devices, biosensors and nanomachines. Here only the gas sensor application is of concern.

7.4.3 Adsorption of Oxygen: Analyses

The most important phenomenon in thin-film sensors takes place at the metal oxide surface. Due to poor coordination between anions and cations in such oxides, which is called natural non-stoichiometry, there exist a number of dangling bonds on the surface, which are effective in adsorbing gaseous species. So there exist a number of gas-sensitive regions around the grain boundary surface (Figure 7.1) [54].

After adsorption, it pulls out electrons from the semiconductor surface due to its strong electron affinity forming a layer of O^- ion on the surface. As electrons are pulled out from the conduction band, a depletion layer of thickness L (also called the Debye length) is thus formed. So now apart from the grain boundary barrier, an electron also encounters the depletion layer as it moves on the surface of the thin film. As a result, a decrease in conductivity is found.

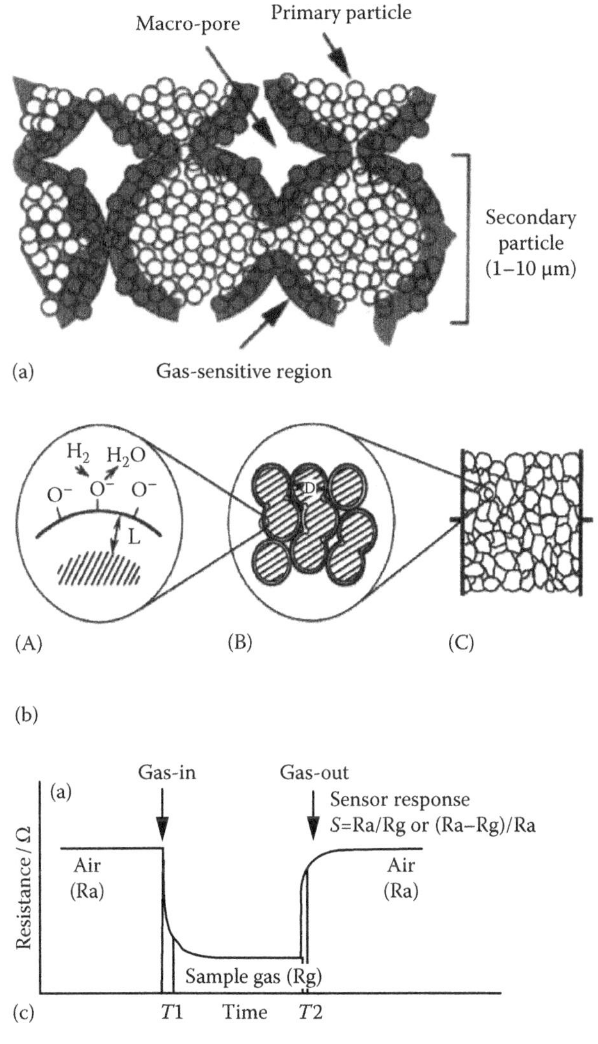

FIGURE 7.1
(a) The gas-sensitive regions on a porous film surface (top view). (b) Oxygen adsorption on a sensor surface: (A) surface (receptor function), (B) microstructure (transducer function) and (C) element (output resistance change). (c) Ideal sensor response to reducing gas. (From Yamazoe, N. et al., *Catal. Surveys from Asia*, 7, 63, 2003.) *Response (*T*1) and recovery (*T*2) time were calculated as the time taken to reach 67% of the saturation value and to reach back to the original value, respectively.

Oxygen adsorption is expressed as follows:

$$O_2 + 2e = 2O^-$$

$$(KO_2PO_2)^{1/2}[e]s = [O^-]$$

where

KO_2 and PO_2 are the adsorption constant and partial pressure of oxygen, respectively

[electrons] and $[O^-]$ are surface densities of free electrons and O^-, respectively

7.4.4 Reaction between Gas (e.g. CH_4) and Oxygen

There is an enormous scope to elaborate this discussion with hundreds of gases, which is beyond our scope. Therefore, we have to stick to a single gas, considering the most concerned toxic and inflammable gas in the present day, i.e. methane. Methane (CH_4) is the simplest alkane (saturated chemical compounds consisting of hydrogen and carbon atoms and are bonded by single bonds without any cycles), the principal component of natural gas and the most abundant organic compound on earth. It is 25 times more effective greenhouse gas than carbon dioxide (CO_2). Methane has a boiling point of −161°C at a pressure of 1 atmosphere. It is flammable only over a narrow range of concentrations (lower explosive limit = 5%, whereas the upper explosive limit = 15%) in air. Liquid methane does not burn unless subjected to high pressure, nearly 4–5 atmospheres.

Now as a reducing gas like hydrogen comes into contact with the surface, it breaks up into H^+ and then reacts with the adsorbed O^- and produces H_2O. In this process, it donates two electrons back to the oxide surface, thus restoring the conductivity. The result of this effect can be easily understood from the sharp increase in conductivity as the gas is exposed. When the gas is removed, the surface is again depleted of carriers as the oxygen is again adsorbed on the surface [55,56]. The effect of this activity in band bending is reflected in Figure 7.2.

Chemisorption occurs in the following steps:

Step-I: $O + e^- \longleftrightarrow O^-$ [150°C–300°C]

Step-II: $2O + e^- \longleftrightarrow O_2^-$ [30°C–150°C]

After the exposure of methane, the reaction becomes

$$CH_4 + 4O^- (ads) \longleftrightarrow CO_2\ (air) + 2H_2O + 4e^- (bulk)$$

As per the reaction, electrons are added to the bulk and there is a reduction in the depletion layer width and hence resistance decreases.

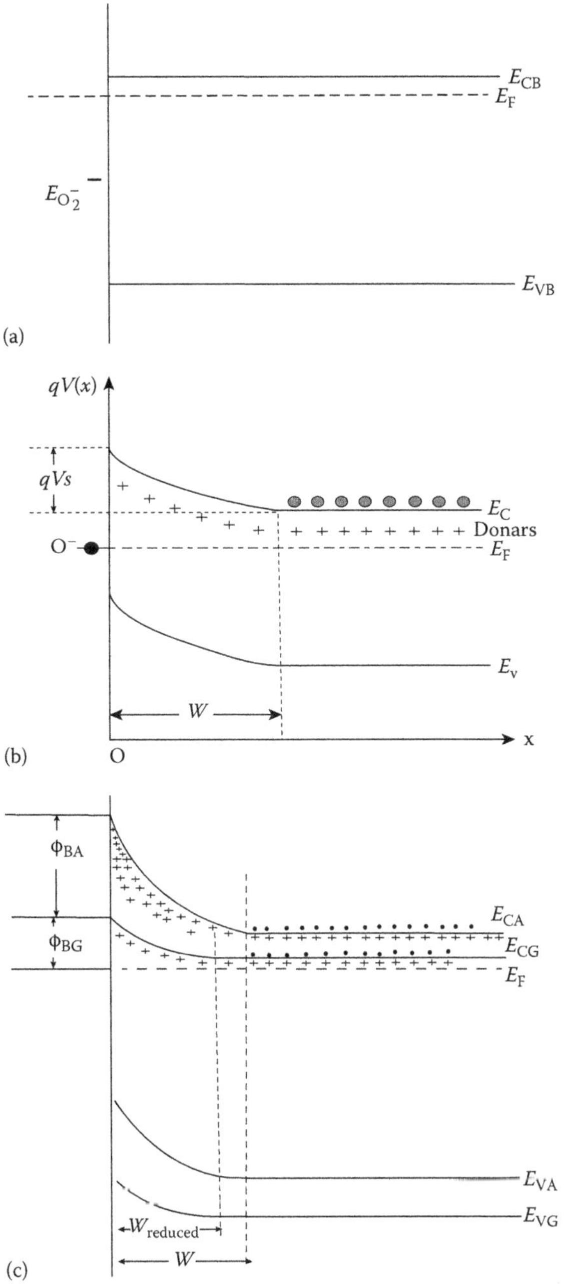

FIGURE 7.2
(a) Energy band diagram at the surface of a semiconducting sensing material before oxygen gas. (b) Energy band diagram after oxygen exposure. (c) Energy band diagram after the exposure of reducing gas. (From Yamazoe, N. and Shimanoe, K., *Sens. Actuators B*, 128, 566, 2008.)

Therefore, by measuring the change in the conductivity of the semiconductor oxide thin films, we can detect the reducing gases. Methane sensing follows the mechanism of hydrogen sensing because hydrogen is produced by the dissociation of methane on noble catalyst metal thin films.

7.4.5 Role of Catalyst on Gas Sensing Mechanism

The effect of noble metals like platinum (Pt), palladium (Pd), ruthenium (Ru), rhodium (Rh), silver (Ag) or gold (Au) on the performance of gas sensors has been registered by many authors, and improved performance of gas sensing has been recorded. These noble metals are also termed as catalytic metals. ZnO-based gas sensors in the form of thin films, with surface modification, are extensively studied to detect the explosive and toxic gases in the surroundings. These noble metals due to its unusual electronic configuration lower the activation energy, thus stimulating the adsorption of the target gas to the metal oxide layer's surface. The noble metals can be included in the device as electrode contact or by surface sensitization. There are two types of sensitization. One is chemical and the other is electronic. In the first part, no charge transfer is taking place, whereas in the latter part, electrons are transferred from the noble metal to the bulk (as shown in Figure 7.3).

As per the fundamentals of Schottky diode, when noble metals are used as electrode contact on the semiconductor surface, there is formation of Schottky contact due to the gap in their work function. Normally, for n-type semiconducting sensing layer, the noble metals should be with higher work function than that of the semiconducting layer, thus increasing the conductance as well as responsivity to target gas by removing the minority charge carrier transportation and decreasing the effect of depletion capacitance and reducing the storage time (t_s). Normally, Pd, Pt, Rh, Ru and Au are used as electrode contact. Lofdahl et al. [57] reported that Pd shows a higher response at the thicker part of the metal thin film, whereas Pt gives more or less the same response for both the thinner and thicker parts of the film.

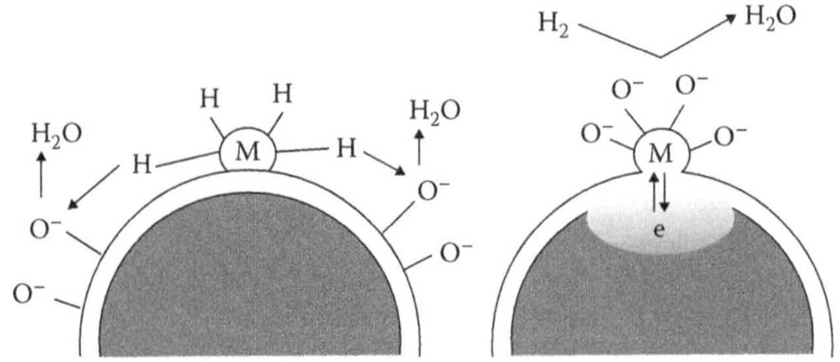

FIGURE 7.3
Chemical and electronic sensitization schematic (left to right). (From Yamazoe, N., *Sens. Actuators B*, 5, 7, 1991.)

But there are some drawbacks in the use of pure Pd metals due to the formation of the β phase of palladium hydride (PdH_2) from the α phase of palladium at low H_2 and at 300 K. To overcome these difficulties, Pd is allowed to a second metal (13%–30% Ag) for H_2 or hydrocarbon sensing. The work function still remains higher than that of an n-type metal oxide. Noble metals like Pt, Pd and Rh are used as surface modifiers also. Surface modification employs several advantages like (1) catalytic activity in the solid–gas interactions and (2) reactivity can be varied by changing the modifier concentration. As the noble metals reduce the activation energy of the reaction, gas sensing occurs at a lower temperature, and as a result, response, response time and recovery time can be improved by noble metal surface modification.

7.4.6 Thick- and Thin-Film Fabrication Process

There are different ways to synthesize the metal oxide layer, either in the form of thick film or in the form of thin film, including chemical vapour deposition (CVD), sputtering, chemical/electrochemical deposition, bulk growth, substrate growth and hydrothermal route. For easy realization, one particular metal oxide layer (e.g. ZnO) has been taken and different growth processes have been discussed.

- Bulk growth – which can be of three types:
  - Melt growth
  - Vapour transport
  - Hydrothermal method
- CVD
  - Substrate growth
  - Chemical route
  - Sputtering

7.4.6.1 Bulk Growth

In spite of various established epitaxial growth techniques, for producing analytical-grade quality material, bulk growth technique is still being adopted by most industries. This can be carried out by three methods: melt growth, vapour phase and hydrothermal.

7.4.6.1.1 Melt Growth

This is one of the versatile methods. The melt is reserved in a pressurized induction melting apparatus. An enclosed cooled crucible is used to hold this melt. As ZnO has been considered, zinc oxide powder is used as the starting material. No thermal energy is applied; rather radio frequency (rf) energy is used as the heat source by employing the induction heating technique. The temperature that is achieved by the induction heating technique is nearly 1900°C, which is sufficient to produce ZnO melt. Once the molten state is achieved, the crucible is detached from the heating chamber to allow crystallization of the melt.

7.4.6.1.2 Vapour Transport

Generally, the vapour phase is used to produce very-high-quality bulk ZnO wafers. In this process, the reaction takes place in a semi-closed horizontal chamber. Pure ZnO powder is used as the ZnO source and is placed at the hot end (~1150°C) of the horizontal chamber. A carrier gas is required as the vapour pressures of O and Zn are lower than ZnO at these temperatures; that's why H_2 is used as a carrier gas to carry the material from one end to the other end maintained at about 1100°C.

The possible reaction in the hot end is

$$ZnO(s) + H_2(g) \rightarrow Zn(g) + H_2O(g)$$

At the cooler end, ZnO is formed by the reverse reaction. To avoid the natural non-stoichiometric nature of ZnO, a minute quantity of water vapour is added.

For low-temperature (nearly 950°C–1000°C) growth technique, chlorine and carbon are used as transporting agents instead of H_2.

7.4.6.1.3 Hydrothermal Method

Hydrothermal synthesis is basically a heterogeneous reaction in the presence of water under high-pressure and high-temperature conditions to dissolve and then recrystallize the material that is insoluble in water under normal conditions. This method needs a seed crystal to start the total procedure.

The crystal growth is performed in a steel vessel (autoclave) having pressure in it, in which nutrients are supplied along with water. The hydrothermal technique is not only suitable for the preparation of monodispersed and highly homogeneous nanoparticles, but it also acts as one of the most efficient techniques for processing nanohybrid and nanocomposite materials.

For the synthesis of ZnO, the hydrothermal method uses ZnO single-crystal seeds (suspended by Pt wire) and moulded ZnO strings. KOH (3 mol/L) and LiOH (1 mol/L) aqueous solutions are used as nutrients. The seeds and the nutrients are positioned in a sealed Pt crucible and kept in an autoclave. The autoclave is then placed into a two-zone vertical furnace. The temperature of the furnace is 300°C–400°C, and the pressure maintained for the growth is between 70 and 100 MPa. ZnO is shifted from the nutrient placed in the high-temperature zone to the seeds in the low-temperature zone. The seeds gradually grow up to form ingots of about 10 mm length roughly after 15 days. To prevent the insertion of impurity from the aqueous solution, another Pt inner container is used. The shape of the crystal depends on the precursor used and the Ph value of the solution and itself on the shapes of seed crystals. In this process, the crystal colour is non-uniform due to the anisotropic crystal growth.

7.4.6.2 Substrate Growth

When used as substrates for epitaxy, proper surface preparation is necessary to evaluate the quality of the grown ZnO. Therefore, in order to reduce the strains and dislocation density in epitaxial ZnO films, lattice mismatch between the substrates should be avoided. As, for example sapphire substrates are preferred for the growth of ZnO using heteroepitaxial growth using a variety of growth techniques. There are so many substrates with minimal lattice mismatch favoured for ZnO growth, e.g. Si, SiC, GaAs, CaF_2 and $ScAlMgO_4$.

One of the popular substrate growth techniques is pulsed laser deposition. In this method, high-power laser pulses are used to evaporate material (ZnO tablet) from a surface known as target. The stoichiometry of the material is not hampered in any way; as a result, a fountain of supersonic particles called plume is directed normal to the surface. The plume fleshes out from the target with different velocity distributions of particles. These particles pack into the substrate, which is preheated and kept at a certain temperature, placed on the opposite side of the target. Single-crystal ZnO has been used to grow high-quality ZnO thin films. The different characteristics of the grown ZnO films depend mainly on the substrate temperature and laser intensity.

7.4.6.3 Chemical Vapour Deposition

CVD is the most popular method because it offers coatings that are conformal, have good step coverage and can coat a large number of wafers at a time. Low-pressure CVD is capable of producing conformal step coverage often with lower electrical resistivity than that from physical vapour deposition.

This technique is very much suitable in the fabrication of epitaxial films also. The reactions take place in a closed chamber where the desired temperature profile is produced using the direction of gas flow. In the CVD method, ZnO deposition occurs as a result of chemical reactions of vapour-phase precursors on the substrate, which are transported into the growth zone by the carrier gas.

7.4.6.4 Sputtering

One of the most popular growth techniques for ZnO deposition is sputtering (dc sputtering, rf magnetron sputtering and reactive sputtering) because of its low cost, simplicity and low operating temperature. Among various types of sputtering, the magnetron sputtering is preferred because sputtering is done from a high-purity ZnO target using an rf magnetron sputter system at a pressure of 10^{-3}–10^{-2} torr in O_2/Ar + O_2 ambient. Ar acts as a catalytic agent and enhances the sputtering technique. Oxygen (O_2) serves as the reactive gas. The rf power applied to the plasma can be varied to regulate the sputtering yield rate from the ZnO target. On the contrary, in dc sputtering, a metallic Zn target is used as a source in an Ar + O_2 gas mixture. For these experiments, an additional step is required, that is presputtering. The target is presputtered for a few minutes before the actual deposition starts to eradicate the contamination present on the target surface to obtain the flawless, optimized ZnO thin film. Of course the physical deposition method has the advantage of producing high-quality thin-film material, but needs a very high temperature to perform. In addition to that, the control over the morphology of the ZnO nanostructures is not simple in this means.

7.4.6.5 Chemical Route

For an extensive use in commercial applications, pure ZnO films must be prepared by a low-temperature deposition methodology. Aqueous chemical synthesis of the nanostructure of ZnO is the most economical and energy-efficient method and enables good morphological control of the nanostructure. In addition, the solution growth technology is suitable to obtain stoichiometric ZnO film because of its oxygen-rich deposition environment, which may be beneficial to the suppression of deep-level-related luminescence and the enhancement of UV emission. In the wet chemical route, the basic building blocks are ions instead of atoms, and therefore, the preparative parameters are easily controllable. It has a number of advantages apart from it being inexpensive, simple and convenient for large-scale deposition: (1) The process can be carried out on any kind of substrate, and (2) unlike the closed vapour deposition method, it does not require high-quality target and/or substrates. Also it does not require vacuum at any stage, (3) the deposition rate and the thickness of the film can be easily controlled by changing the deposition cycles, and (4) it is a low-temperature chemical solution method and does not cause local overheating that can be detrimental for materials to be deposited.

Nanocrystalline ZnO particles can be prepared by the aqueous solution of zinc salt ($ZnSO_4$, $ZnNO_3$, etc.) without any requirement of the calcination step at high temperature. An aqueous solution of zinc salt and sodium hydroxide are mixed and stirred using a magnetic stirrer. The solution is kept aside and the crystal clear homogeneous solution is separated. Either by dipping or dropping technique, the ZnO thin film has been prepared with nanocrystals of different sizes and shapes.

7.4.7 Sensor Characterizations

7.4.7.1 X-Ray Diffraction

X-ray diffraction (XRD) is a dominant tool for material characterization as well as for detailed structural explanation. As the physical properties of solids (e.g. electrical, optical, magnetic) depend on atomic arrangements of materials, determination of the crystal structure is a crucial part of the structural and chemical characterization of materials. X-ray patterns are used to establish the atomic arrangements of the materials because of the fact that the lattice parameter, *d* (spacing between different planes), is of the order of x-ray wavelength. Further, the XRD method can be used to distinguish crystalline materials from nanocrystalline (amorphous) materials. Structure identification is made from XRD pattern analysis, comparing it with the internationally recognized database containing the reference pattern (JCPDS). From the XRD pattern, we can obtain the following information:

1. To judge formation of a particular material system.
2. Unit cell structure, lattice parameters, and miller indices.
3. Types of phases present in the material.
4. Estimation of crystalline/amorphous content in the sample.
5. Evaluation of the average crystalline size from the width of the peak in a particular phase pattern. Large crystal size gives rise to sharp peaks, while the peak width increases with decreasing crystal size.
6. An analysis of structural distortion arising as a result of variation in d-spacing caused by the strain and thermal distortion.

7.4.7.2 Determination of Crystal Size

XRD analysis has been the most popular method for the estimation of crystallite size in nanomaterials and, therefore, has been extensively used in the present work. The evaluation of crystallite sizes in the nanometre range warrants careful analytical skills. The broadening of the Bragg peaks is ascribed to the development of the crystallite refinement and internal stain. For size broadening and strain broadening, the full width at half maximum (FWHM) of the Brag peaks as a function of the diffraction angle is analysed. Crystallite size of the deposits is calculated by the XRD peak broadening. The diffraction patterns are obtained using Cu Kα radiation at a scan rate of 10/min. The FWHM of the diffraction peaks were estimated by pseudo-Voigt curve fitting. After subtracting the instrumental line broadening, which was estimated using quartz and silicon standards, the grain size can be estimated by the Scherrer equation $D = .9\lambda/\mathrm{B}\cos\theta$, where λ is the wavelength of the x-ray, β is FWHM in radian and θ is the peak angle.

7.4.7.3 Field Emission Scanning Electron Microscopy

This is one of the tools to find out the material morphology. In this technique, electrons are used instead of light waves to see the microstructure of the surface of a specimen. However, since electrons are excited to high energy (keV), the wavelength of electron waves is quite small and resolution is quite high. In a field emission scanning electron microscopy, electrons are thermionically emitted from a tungsten cathode and are accelerated towards the anode. Tungsten is used because it has the highest melting point and lowest vapour pressure. The electron beam, which typically has an energy ranging from a few hundred eV to 100 keV, is focused by one or two condenser lenses into a beam with a very fine focal spot sized 1–5 nm. The beam passes through pairs of scanning coils and falls on the objective lens, which deflects the beam horizontally and vertically to scan in a raster fashion to cover a rectangular area of the sample. The size of the interaction volume depends on the beam accelerating voltage, the atomic number of the specimen and the specimen's density. The energy exchange between the electron beam and the sample results in electromagnetic radiation, which is used to produce an image.

Backscattered electrons are also used to detect contrast between areas with different chemical compositions.

7.4.7.4 Transmission Electron Microscopy

Transmission electron microscopy (TEM) has become a central part in the inventory of characterization techniques of materials. TEM's positive point is its high lateral spatial resolution and its capacity to provide both image and diffraction information from a single sample. Here, the highly energetic beam of electrons used in TEM interacts with the sample to produce characteristic radiation. In TEM, a focused electron beam is incident on a thin (less than 200 nm) sample. The signal in TEM is obtained from both undeflected and deflected electrons from the surface. A series of magnetic lenses at and below the sample position are responsible for handing over the signal to a detector. There is a magnification of the spatial information. This magnification range is facilitated by the small wavelength of the incident electrons and is key to the unique capabilities associated with TEM analysis. TEM offers two methods of observation: diffraction mode and image mode. In diffraction mode, an electron diffraction pattern is obtained on the fluorescent screen. The diffraction pattern is entirely equivalent to an XRD pattern. It can alter between image and diffraction modes depending on the requirement. The reasons for this simplicity are buried in the intricate electron optics technology that makes the practice of TEM possible.

There are a number of drawbacks to the TEM technique. Many materials require extensive sample preparation to produce a thin sample which makes TEM analysis a time-consuming process with low throughput.

7.4.7.5 Photoluminescence Spectroscopy

Photoluminescence (PL) deals with the spontaneous emission of light from a material under optical excitation. The excitation energy and intensity are chosen to point different regions and excitation concentrations in the sample. PL investigations can be used to characterize a variety of material parameters such as to identify surface, interface and impurity levels and to measure alloy disorder and interface roughness. The intensity of the PL signal provides information on the quality of surfaces and interfaces. PL intensity can be varied under an applied bias and can be used to map the electrical field at the surface of a sample. Also, PL intensity varies with temperature. PL analysis is non-destructive, which means that the sample can be reused. No electrical contacts are required. In addition, time-resolved PL can be very swift; it is to characterize the most rapid processes in a material. The main constraint of PL analysis is its radiative events. Materials with poor radiation, i.e. low-quality indirect band gap semiconductors, are not easy to study via ordinary PL. To understand PL effectively, when light of sufficient energy is incident on a material, photons are absorbed and electrons are excited. When these excitations come to its original state, radiative relaxation occurs; the emitted light is called PL. This light is collected and analysed to get the information about the photo-excited material. The PL spectrum can be used also to determine electronic energy levels. Moreover, the PL intensity gives a measure of the relative rates of radiative and non-radiative recombination. PL intensity may vary with external parameters like temperature and applied voltage, and this variation can be used to characterize further the underlying electronic states and bands. Finally, compared with other optical methods of characterization, PL is less stringent about beam alignment, surface flatness and sample thickness.

7.4.7.6 Fourier Transform Infrared Spectroscopy

Fourier transform infrared spectroscopy (FTIR) is most useful for identifying chemicals that are either organic or inorganic. It can be utilized to get information of some components of an unknown mixture. It can be applied to the analysis of solids, liquids and gases. The term Fourier transform infrared spectroscopy refers to a fairly recent development in the manner in which the data are collected and converted from an interference pattern to a spectrum. Today's FTIR instruments are computerized, which makes them faster and more sensitive than the older dispersive instruments.

7.4.7.7 Qualitative Analysis

FTIR can be used to identify chemicals from spills, paints, polymers, coatings, drugs and contaminants. FTIR is perhaps the most powerful tool for identifying types of chemical bonds (functional groups). The wavelength of light absorbed is characteristic of the chemical bond as can be seen in this annotated spectrum. By interpreting the infrared (IR) absorption spectrum, the chemical bonds in a molecule can be determined. The energy corresponding to these transitions between molecular vibrational states is generally 1–10 kcal/mole, which corresponds to the IR portion of the electromagnetic spectrum.

7.4.7.8 UV/VIS Spectroscopy

Ultraviolet–visible spectroscopy (UV/VIS) involves the spectroscopy of photons in the UV/VIS region. It uses light in the VIS and adjacent near-UV and near-IR ranges. In this region of the electromagnetic spectrum, molecules undergo electronic transitions. UV/VIS spectroscopy is routinely used in the quantitative determination of solutions of transition metal ions and highly conjugated organic compounds.

Solutions of transition metal ions can be coloured (i.e. absorb visible light) because d electrons within the metal atoms can be excited from one electronic state to another. The colour of metal ion solutions is strongly affected by the presence of other species, such as certain anions. For instance, the colour of a dilute solution of copper sulphate is light blue; adding ammonia intensifies the colour and changes the wavelength of maximum absorption (λ_{max}).

Organic compounds, especially those with a high degree of conjugation, also absorb light in the UV or VIS regions of the electromagnetic spectrum. The solvents for these determinations are often water for water-soluble compounds, or ethanol for organic-soluble compounds. Solvent polarity and pH can affect the absorption spectrum of an organic compound. While charge transfer complexes also give rise to colours, the colours are often too intense to be used for quantitative measurement.

The Beer–Lambert law states that the absorbance of a solution is directly proportional to the solution's concentration. Thus, UV/VIS spectroscopy can be used to determine the concentration of a solution. A UV/VIS spectrophotometre may be used as a detector. The presence of an analyte gives a response which can be assumed to be proportional to the concentration. The height of the peak for a particular concentration is known as the response factor. The method is most often used in a quantitative way to determine concentrations of an absorbing species in solution, using the Beer–Lambert law:

$$A = -\log 10(I/I_0) = \varepsilon.\ c.\ L$$

where

A is the measured absorbance
I_0 is the intensity of the incident light at a given wavelength
I is the transmitted intensity
L the path length through the sample
c is the concentration of the absorbing species
For each species and wavelength,
ε is a constant known as the molar absorptivity

7.4.7.9 Raman Spectroscopy

Raman spectroscopy provides information about molecular vibrations that can be used for sample identification and quantitation. The technique involves incidenting a monochromatic light source (i.e. laser) on a sample and detecting the scattered light. The majority of the scattered light is of the same frequency as the excitation source; this is known as Rayleigh or elastic scattering. A very small amount of the scattered lights shifted in energy from the laser frequency due to interactions between the incident electromagnetic waves and the vibrational energy levels of the molecules in the sample. Plotting the intensity of this *shifted* light versus frequency results in a Raman spectrum of the sample.

Raman spectroscopy can be used for both qualitative and quantitative applications. Raman spectra are very specific, and chemical identifications can be performed by using search algorithms against digital databases. As in infrared spectroscopy, band areas are proportional to concentration, making Raman essential to quantitative analysis.

- The spectral range starts from below 400 cm^{-1}, making the technique ideal for both organic and inorganic species.
- Raman spectroscopy can be used to measure bands of symmetric linkages (e.g. –S–S–, –C–S–, –C=C–) which cannot be measured with other characterization techniques.

7.4.8 Sensor Reliability Issues

The ability of a sensor to maintain its performance characteristics for a certain period of time is determined by a reliability study. Unless otherwise stated, it is the ability of a sensor to reproduce output readings, obtained during calibration for a specific period of time. There should not be much drift after many cycles of operation for good stability.

The fabrication of a sensor structure does not confer the thermal stability of a microheater as well as the stability of a response magnitude. For the thermal stability of a microheater, the calibration of temperature coefficient of resistance with respect to temperature is an essential part; otherwise, a sharp rise in resistance with temperature will give an erroneous result. The variation in microheater resistance from room temperature to required temperature should be very small thereby establishing the stable high-temperature performance of the microheater with negligible shift from the baseline value.

Generally, the stability study of the sensor is carried out in an inert atmosphere for several hours with and without the presence of the target gas. The linearity of the graph signifies the stability of the sensor.

References

1. H. Kim and W. Sigmund, ZnO nanocrystals synthesized by physical vapor deposition, *Nanotechnology* 4, 275–278 (2004).

2. X. D. Wang, C. J. Summers and Z. L. Wang, Large-scale hexagonal-patterned growth of aligned ZnO nanorods for nano-optoelectronics and nanosensor arrays, *Nano Lett.* 4, 423–427 (2003).
3. S. Dalui, S. N. Das, R. K. Roy, R. N. Gayen and A. K. Pal, Synthesis of DLC films with different sp^2/sp^3 ratios and their hydrophobic behaviour, *Thin Solid Films* 5, 516–519 (2008).
4. X. Y. Kong and Z. L. Wang, Spontaneous polarization-induced nanohelixes, nanosprings, and nanorings of piezoelectric nanobelts, *Nano Lett.* 3, 1625 (2003).
5. X. D. Wang, C. J. Summers and Z. L. Wang, Mesoporous single crystal ZnO nanowires epitaxially sheathed with Zn_2SiO_4, *Adv. Mater.* 16, 1215 (2004).
6. L. Shi, Q. Hao, C. Yu, N. Mingo, X. Kong and Z. L. Wang, Thermal conductivities of individual tin dioxide nanobelts, *Appl. Phys. Lett.* 84, 2638 (2004).
7. Z. L. Wang, Zinc oxide nanostructures: Growth, properties and applications, *J. Phys. Condens. Matter* 16, 829 (2004).
8. P. Bhattacharyya, P. K. Basu, H. Saha and S. Basu, Fast response methane sensor based on Pd(Ag)/ZnO/Zn MIM structure, *Sensors Lett.* 4, 371–376 (2006).
9. P. K. Basu, P. Bhattacharyya, N. Saha, H. Saha and S. Basu, Methane sensing properties of platinum catalysed nano porous zinc oxide thin films derived by electrochemical anodization, *Sensors Lett.* 6, 219–225 (2008).
10. S. J. Fonash, J. A. Roger and C. H. S. Dupuy, AC equivalent circuits for MIM structures, *J. Appl. Phys.* 45(7), 2907–2910 (1974).
11. C. L. Wu, L. Chang, H. G. Chen, C. W. Lin, T. F. Chang, Y. C. Chao and J. K. Yan, Growth and characterization of chemical-vapor-deposited zinc oxide nanorods, *Thin Solid Films* 498, 137–141 (2006).
12. H. Yu, Z. Zhang, M. Han, X. Hao and F. Zhu, A general low-temperature route for large-scale fabrication of highly oriented ZnO nanorod/nanotube arrays, *J. Am. Chem. Soc.* 127, 2378–2379 (2005).
13. X. Y. Kong, Y. Ding, R. Yang and Z. L. Wang, Single-crystal nanorings formed by epitaxial self-coiling of polar nanobelts, *Science* 303, 1348–1351 (2004).
14. D. M. Bagnall, Y. F. Chen, Z. Zhu, T. Yao, S. Koyama, M. Y. Shen and T. Goto, Optically pumped lasing of ZnO at room temperature, *Appl. Phys. Lett.* 70, 2230–2232 (1997).
15. K. Balachandra Kumar and P. Raji, Synthesis and characterisation of nano Zinc Oxide by Sol Gel spin coating II, *Recent Res. Sci. Technol.* 3, 48–52 (2011).
16. A. Mondal, N. Mukherjee and S. K. Bhar, Galvanic deposition of hexagonal ZnO thin films on TCO glass substrate, *Mater. Lett.* 60, 1748–1752 (2006).
17. K. E. Peterson, Silicon as a mechanical material, *IEEE Proc.* 70, (1982).
18. W. P. Eaton, J. H. Smith, D. J. Monk, G. O. Brien and T. F. Miller, Comparison of bulk- and surface-micromachined pressure sensors micromachined devices and components, *Proc. SPIE* 3514, 431–434 (1998).
19. C. Liu, *Foundation of MEMS*, Pearson Education, Inc., New Jersey. Chapter-10, 11 (2006).
20. K. E. Bean, Anisotropic etching of silicon, *IEEE Trans.* ED-25, 10, 1185 (1978).
21. H. Y. Chu and W. Fang, A novel convex corner compensation for wet anisotropic etching on (100) Silicon wafer, *Proc. IEEE* 253–256 (2004).
22. D. B. Lee, Anisotropic etching of silicon, *J. Appl. Phys.* 40, 4569–4574 (1969).
23. M. Bustillo, R. T. Howe and R. S. Muller, Surface micromachining, *J. Microelectromech. Syst.*, 86, 1552–1574 (1998).

24. A. Salehi, A. Nikfarjam and D. J. Kalantari, Highly sensitive humidity sensor using Pd/porous GaAs Schottky contact, *IEEE Sens. J.* 6, 1415–1421 (2006).
25. N. Tasaltin, F. Dumludag, M. A. Ebeoglu, H. Yüzer and Z. Z. Ozturk, Pd/native nitride/nGaAs structures as hydrogen sensors, *Sens. Actuators B: Chem.* 130, 59–64 (2008).
26. S. K. Hazra and S. Basu, Hydrogen sensitivity of ZnO p–n homojunctions, *Sens. Actuators B: Chem.* 117, 117–182 (2006).
27. R. C. Hughes, W. K. Schubert, T. E. Zipperian, J. L. Rodriguez and T. A. Plut, Thin film palladium and shiver alloys and layers for metal insulator semiconductor sensors, *J. Appl. Phys.* 62, 1074–1082 (1987).
27. A. N. Banerjee and K. K. Chattopadhyay, Progress in crystal growth and characterization of materials, 50, 52–105 (2005).
28. S. Xu and Z. L. Wang, One-dimensional ZnO nanostructures: Solution growth and functional properties, *Nano Res.* 4, 1013–1098 (2011).
29. Y. Hu, X. Zhou, Q. Han, Q. Cao and Y. Huang, Sensing properties of CuO-ZnO heterojunction gas sensors, *Mater. Sci. Eng. B* 99, 41–43 (2003).
30. Z. Ling, C. Leach, and R. Freer, Heterojunction gas sensors for environmental NO_2 and CO_2 monitoring, *J. Eur. Ceramic Soc.* 21, 1977–1980 (2001).
31. P. Bhattacharyya, G. P. Mishra, S. K. Sarkar, The effect of surface modification and catalytic metal contact on methane sensing performance of nano-ZnO-Si heterojunction sensor, *Microelectronics Reliability* 51, 2185–2194 (2011).
32. Z. Ling and C. Leach, The effect of relative humidity on the NO_2 sensitivity of a SnO_2/WO_3 heterojunction gas sensor, *Sens. Actuators B: Chem.* 102, 102–106 (2004).
33. Y. Xue-Jun, H. Tian-Sheng, X. Wei, C. Kun and X. Xing, High performance micro CO sensors based on ZnO-SnO_2 composite nanofibers with anti-humidity characteristics, *Chin. Phys. Lett.* 29, 120702 (2012).
34. M. Radecka, K. Zakrzewska and M. Rekas, SnO_2-TiO_2 solid solutions for gas sensors, *Sens. Actuators B* 47, 194–204 (1998).
35. L. E. Depero, M. Ferroni, V. Guidi et al., Preparation and micro-structural characterization of nanosized thin film of TiO_2-WO_3 as a novel material with high sensitivity towards NO_2, *Sens. Actuators B* 36, 381–383 (1996).
36. J. S. Suehle, R. E. Cavicchi, M. Gaitan and S. Semancik, Tin oxide gas sensor fabricated using CMOS micro-hotplates and in-situ processing, *IEEE Electron Dev. Lett.* 14, 118–120 (1993).
37. S. Semancik, R. E. Cavicchi, K. G. Kreider, J. S. Suehle, P. Chaparala, Deposition of multiple active films for conductometric microsensor arrays, *Proceedings of the Transducers, Eurosensors IX*, Stockholm, Sweden, vol. 95, 1995, pp. 831–834.
38. S. Roy, T. Majhi, A. Kundu, C. K. Sarkar and H. Saha, Design, fabrication and simulation of coplanar microheater using nickel alloy for low temperature gas sensor application, *Sensor Lett.* (ASP) 9, 1382–1389 (2011).
39. R. Cavicchi, J. Suehle, P. Chaparala, K. Kreider, M. Gaitan, and S. Semancik, Microhotplate gas sensor, *Proceedings of the 1994 Solid State Sensor and Actuator Workshop*, Hilton Head, SC, pp. 53–56 (1994).
40. J. Puigcorbe, A. Vila, J. Cerda, A. Cirera, I. Gracia, C. Cane and J. R. Morante, Thermo-mechanical analysis of a microdrop coated gas sensor, *Sens. Actuators A* 97, 379–385 (2000).

41. J. Gong, Q. Chen, W. Fei and S. Seal, Micromachined nanocrystalline SnO_2 chemical gas sensors for electronic nose, *Sens. Actuators B* 102, 117–125 (2004).
42. I. J. Kim, S. D. Han, C.H. Han, J. Gwak and D. U. Hong, Development of micro hydrogen gas sensor with SnO_2–Ag_2O–PtOx composite using MEMS process, *Sens. Actuators B* 127, 441–446 (2007).
43. P.K. Basu, P. Bhattacharyya, N. Saha, H. Saha and S. Basu, Methane sensing properties of platinum catalysed nano porous Zinc Oxide thin films derived by electrochemical anodization, *Sensor Lett.* 6, 219–225 (2008).
44. C. Wang, L. Yin, L. Zhang, D. Xiang and R. Gao, Metal oxide gas sensors: Sensitivity and influencing factors, *Sensors* 10, 2088–2106 (2010).
45. P. Bhattacharyya, P. K. Basu, H. Saha and S. Basu, Noble metal catalytic contacts to sol-gel nanocrystalline zinc oxide thin films for sensing methane, *Sens. Actuators B*, 129, 551–557 (2008).
46. T. Hyodo, Y. Baba, K. Wada, Y. Shimizu and M. Egashira, Hydrogen sensing properties of SnO_2 varistors loaded with SiO_2 by surface chemical modification with diethoxy-dimethylsilane, *Sens. Actuators B* 64, 175–181 (2000).
47. J. Li, Y. J. Lu, Q. Ye, M. Cinke, J. Han and M. Meyyappan, Room temperature methane detection using palladium loaded single-walled carbon nanotube sensors, *Chem. Phys. Lett.* 391, 344–348 (2004).
48. W. Brattain and J. Bardeen, Surface properties of germanium, *Bell Telephone Syst. Tech. Publs.* 2086, 1–41 (1953).
49. T. Seiyama, A. Kato, K. Fujushi and M. Nagatani, A new detector for gaseous component using semiconductive thin films, *Anal. Chem.* 34, 1502 (1962).
50. P. de Moor, A. Witvrouw, V. Simons and I. de Wolf, The fabrication and reliability testing of Ti/TiN heaters, *Proc. SPIE*, Santa Clara, California, 3874, 284–293 (1999).
51. K. Zakrzewska, K. Zakrzewska, *Thin Solid Films* 391, 229–238 (2001).
52. S. Basu and P. K. Basu, Review article nanocrystalline metal oxides for methane sensors: Role of noble metals, *J. Sensors*, 861968, 20 (2009).
53. E. Comini, M. Ferroni, V. Guidi, G. Faglia, G. Martinelli and G. Sberveglieri, Nanostructured mixed oxides compounds for gas sensing applications, *Sens. Actuators B* 84, 26–32 (2002).
54. N. Yamazoe, G. Sakai and K. Shimanoe, Oxide semiconductor gas sensors, *Catal. Surveys from Asia* 7, 63–75 (2003).
55. E. Jones, The pellistor catalytic gas selector, in *Solid State Gas Sensors*, Ed. P. T. Moseley and B. C. Tofield, Adam Hilger, Bristol, U.K., 1987, pp. 17–31.
56. A. Janotti and C. G. V. de Walle, Fundamentals of zinc oxide as a semiconductor, *Reports Progress Phys.* 72, 126501 (2009).
57. M. Lofdahl, C. Utaiwasin, A. Carlsson, I. Lundstrom and M. Eriksson, Gas response dependence on gate metal morphology of field effect devices, *Sens. Actuators B* 80, 183–192 (2001).
58. N. Yamazoe and K. Shimanoe, Theory of power laws for semiconductor gas sensors, *Sens. Actuators B*, 128, 566–573 (2008).
59. N. Yamazoe, New approaches for improving semiconductor gas sensors, *Sens. Actuators* B, 5, 7–19 (1991).

8

Sensing with Graphene

8.1 Introduction

The monolayer 'graphene' was first invented in the year 1986 and was presented as a 2D sheet of perfect honeycomb lattice of single carbon atoms, which was first formed by graphite intercalation [1]. Carbon being the centre of attention of many research fields has an exceptional electronic configuration that allows it to form different types of hybridized atomic orbitals. Carbon atoms in the elemental substances (e.g. diamond, graphite) make covalent bonds to each other. This highly directive covalent bond facilitates the carbon atom to adapt into various molecular and crystalline structures. This ultimately settles the varied chemical properties and physical properties of the carbon allotropes [2].

Carbon nanotube–based devices have also proven record in the various fields of electronics. Still, it deals with some negative issues, for example CNT has 1D structure that is not appropriate to be used in existing electronic devices as the fabrication on top of CNT becomes harder. On the contrary, graphene [3] has got 2D structure of one-atom thick carbon. It is easier to fabricate the next layer on top of graphene.

One description given in a recent analysis on graphene is as follows:

> Graphene is a flat monolayer of carbon atoms tightly packed into a two-dimensional (2D) honeycomb lattice, and is a basic building block for graphitic materials of all other dimensionalities. It can be wrapped up into 0D fullerenes, rolled into 1D nanotubes or stacked into 3D graphite [4].

Figure 8.1 shows different carbon allotropes.

As there is no obstruction for the electrons inside the graphene layer, electrons behave there as if they have no mass, travelling at a speed of thousand kilometres per second. Due to the unique electronic structure of carbon atom, it can form many different types of bonds and hence can be found in different forms in the nature. The four valence electrons of carbon atom distribute themselves as $1s^2 2s^2 2p_x^1 2p_y^1$.

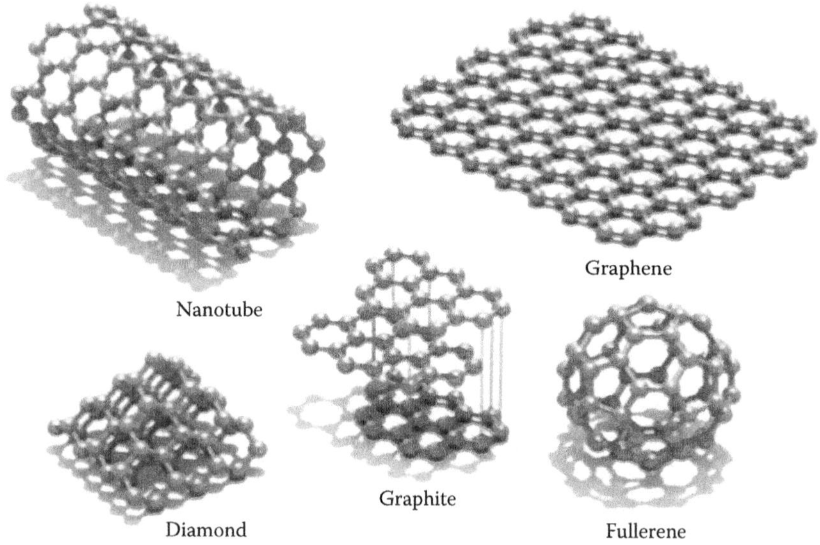

FIGURE 8.1
Carbon allotropes: fullerene (0D), carbon nanotube (1D) and graphene (2D).

The 2s and 2p orbitals are very close to each other, and eventually, an electron has been shifted from the 2s orbital to the empty 2p orbital to offer four unpaired electrons. This is a very preferred phenomenon as all four orbitals become exactly half filled and can be used to form strong covalent bonds, which consequently reduces the overall excitation energy to start a reaction. Figure 8.2 shows the electron distribution in sp^2 hybridization.

The valence orbitals can make hybridization in three different ways. The first one is sp hybridization, the second one is sp^2 hybridization and the third one is sp^3 hybridization, as shown in Figure 8.3. First, in sp hybridization, one 2s orbital and one of the 2p orbitals hybridize to form two sp orbitals forming two symmetrical covalent sigma (σ) bonds with a bond angle of 180°. The two remaining electrons in the $2p_y$ and $2p_z$ orbital form pi(Π) bonds perpendicular to the σ bonds. When two sp hybridized carbons are close to each

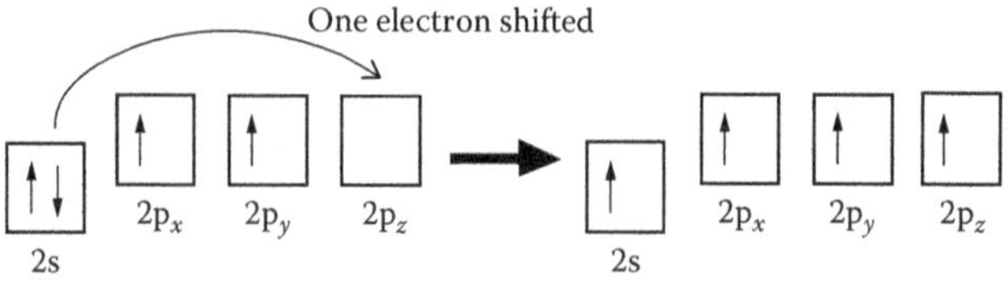

FIGURE 8.2
Electron distribution in sp^2 hybridization.

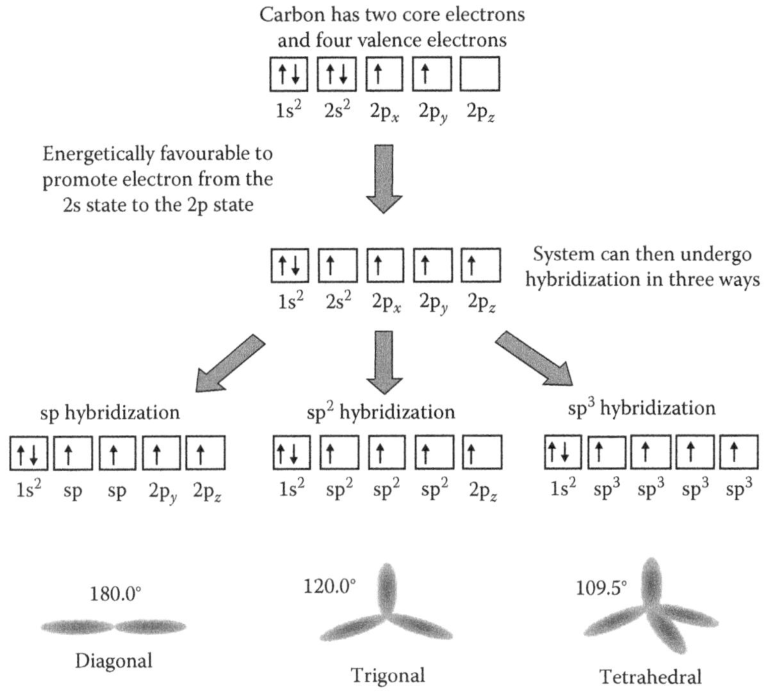

FIGURE 8.3
Schematic of different hybridizations.

other, the π orbitals surround the σ bond and form a triple bond. This occurs in the alkyne family of hydrocarbons.

Next, in sp^3 hybridization, all the four orbitals can hybridize to form four new sigma (σ) bonds that are symmetrical. This forms a tetrahedral structure with a bond angle of 109.5°. Being associated with σ bond, each valence electron is tightly bound within the structure rendering all the electrons to be localized to an atom, thus making it electrically insulating. The antibonding state σ* is at such a high energy level that electrons are not able to reach this state upon excitation. sp^3 hybridization is observed in diamond.

The carbon atom doesn't have enough unpaired electrons to form the required number of bonds, so it needs to promote one of the $2s^2$ pair into the empty $2p_z$ orbital. In case of sp^2 hybridization, with a small excitation, an electron in the 2s subshell can be promoted into the 2p subshell. This time, the carbon atom hybridizes with three of the orbitals. One s orbital and two p orbitals directed to each other now join together to give molecular orbitals, each containing a bonding pair of electrons. These are sigma bonds. The p_z orbitals are so close that the clouds of electrons overlap sideways above and below the plane of the molecule. A bond formed in this way is called a pi bond that is a very weak bond. The sigma bond as well as pi bond lay in the same plane, where the p orbital is at right angles to the s orbital. Any twist in the molecule would break the weak pi bond. In case of graphene, the three sp^2 hybrid orbitals arrange themselves at 120° to each other in a plane.

Before starting the detailed study, one should be aware of the different parameters associated with graphene. The distance between carbon atoms in graphene sheet is 0.142 nm. This hexagonally placed carbon atom of graphene has an area of 0.052 nm^2. The graphene film density is approximately 1.5–2.0 g/cm^3. Graphene is almost transparent; it absorbs only 2.3% of the light intensity, independent of the wavelength in the optical domain. This number is given by $\pi\alpha$, where α is the fine structure constant. Intrinsic graphene has a breaking strength of 42 N/m. The hypothetical 2D steel has a breaking strength of 0.084–0.40 N/m. Thus, the mechanical strength of graphene is more than 100 times greater than steel, making it the strongest material in the world. Using the layer thickness, we get a bulk conductivity of $0.96 \times 10^6\ \Omega^{-1}\ cm^{-1}$ for graphene. Using the layer thickness, we get a bulk conductivity of $0.96 \times 10^6\ \Omega^{-1}\ cm^{-1}$ for graphene, which is higher than the conductivity of copper ($0.60 \times 10^6\ \Omega^{-1}\ cm^{-1}$). The thermal conductivity of graphene is approximately 5000 W/m/K. The thermal conductivity of copper at room temperature is 401 W/m/K. The thermal conductivity of graphene is approximately 5000 W/m/K which is 10 times higher than that of copper (401 W/m/K). The melting point of graphene is estimated to be about 6560°F, one of the highest known for any material. No boiling point has yet been established.

8.2 Properties of Graphene

Graphene consists of one s-orbital and two in-plane p-orbitals that undergo sp^2 hybridization. The shape of the hybridized orbital is trigonal wherein each carbon atom in graphene has three nearest carbon atoms making three in-plane σ bonds per atom. These σ bonds are very strong compared to pi bond due to which graphene possesses an unbreakable hexagonal structure. The last p-orbital overlaps with each other from surrounding carbon atoms laterally and forms pi-bonds that exist above and below each atomic layer. The σ-electrons are tightly bound to the atom and therefore are not responsible anyway for the conduction phenomenon, while in large sheets the π and π^* orbitals become valence and conduction bands, which dominate the planar conduction phenomena.

As Figure 8.4 shows, the hexagonal carbon lattice can be subdivided into basic identical trigonal sublattices. An atom in sublattice A is bonded to three atoms from sublattice B and vice versa. The orientation of carbon atoms on the edges of the graphene layer determines whether it is armchair or zigzag. Zigzag contributes to the metallic behaviour of a substance, whereas armchair contributes to semiconducting behaviour.

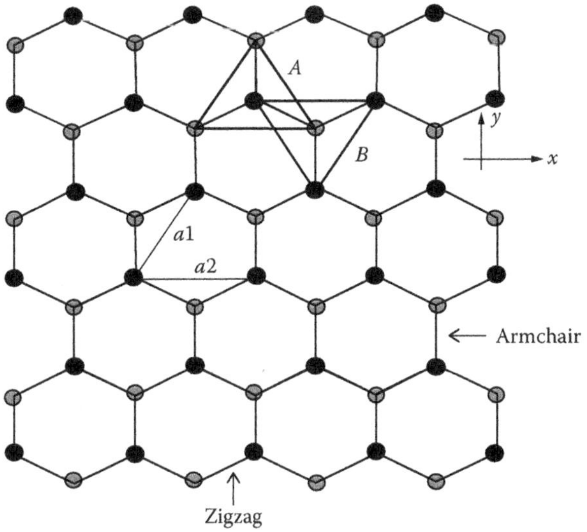

FIGURE 8.4
The hexagonal graphene lattice displaying both zigzag and armchair edges is made up of two identical sublattices, *A* and *B*.

8.2.1 Electronic Property

The electronic property of graphene depends on the electronic structure as it is differentiable from other 3D structures. As electrons hop between sublattices, they produce an effective magnetic field proportional to their momentum. Because of their symmetry, the momentum-dependent field vanishes at certain high symmetry points, known as Dirac points (shown in Figure 8.5) where the conduction and the valence band form conically shaped valleys

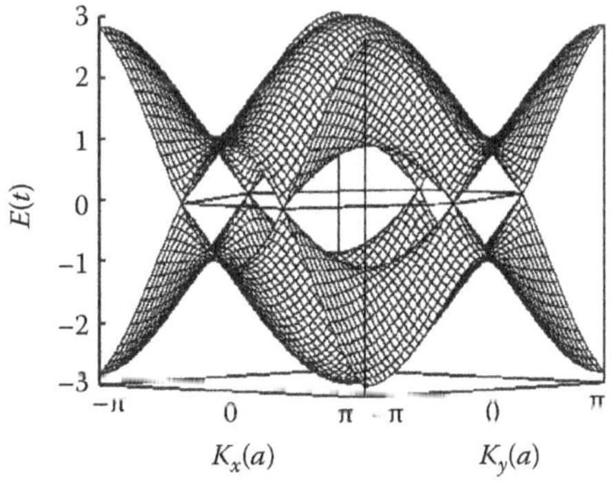

FIGURE 8.5
Energy spectrum of the energy bands close to one of the Dirac points.

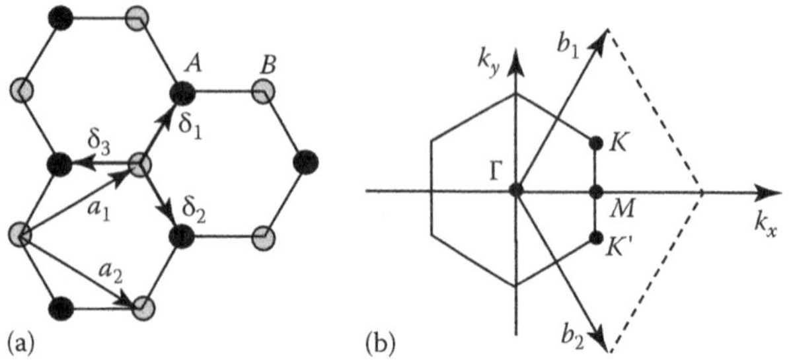

FIGURE 8.6
(a) Honeycomb lattice structure of graphene, made out of two interpenetrating triangular lattices (*a*1 and *a*2 are the lattice unit vectors, and *i*, *i* = 1, 2, 3 are the nearest-neighbour vectors and its Brillouin zone). (b) Corresponding Brillouin zone. The Dirac cones are located at the *K* and *K′* points.

that touch at the six corners of the Brillouin zone (shown in Figure 8.6) making graphene a zero-gap material with respect to conduction in those directions. Depending on the definition, the effective mass (m^*) [5] of an electron is the second derivative of energy with respect to the electron's momentum vector. We know the second derivative of any linear function is zero, so electrons in graphene have zero effective mass near the Dirac points. In high symmetry lattice directions, graphene is a zero-gap semiconductor in which massless electrons travel at constant speed [6–8]. Actually, the electron effective mass represents how electrons move in an applied field.

The Fermi level passes through the K or Dirac points. The precise position of the Dirac point and Fermi level at the graphene/oxide interface has yet to be investigated. Xu et al. [9] reported the first direct measurement of the Dirac point, the Fermi level and the work function of single-layer graphene (SLG) by using photoemission threshold spectroscopy.

In a pure graphene sheet, the Fermi level is positioned at the converging points of these cones. The density of states is zero at that converging point; the electrical conductivity of pure graphene is quite low. Just like a usual semiconductor, the position of the Fermi level is also changed by some externally applied energy and becomes either n-doped (with electrons) or p-doped (with holes). Similarly, graphene can also be doped by three different ways. First, the heteroatom doping, including arc discharge, chemical vapour deposition (CVD), electrothermal reaction and ion-irradiation approaches; second, the chemical modification strategy; third, the method of electrostatic field tuning. The electrical conductivity is obviously high for doped graphene, which may be higher than copper.

8.2.2 Mechanical Properties

Graphene is one of the strongest materials known with a breaking strength over 100 times greater than a steel film. The mechanical properties should be calculated as they may diverge from the well-known properties of graphite. The carbon–carbon bond length in graphene is about 0.142 nm. We can accurately measure the length, width and thickness of the suspended graphene sheet and also the spring constant and many others by atomic force microscopy (AFM). Van der Waals forces are mainly responsible for making the carbon atom tightly bound to each other to make a graphene sheet, where an AFM tip was probed to test its mechanical properties. The spring constant of a monolayer graphene is 1–5 N/m and Young's modulus is 1.0 TPa for an ideal graphene sheet which may vary maximum up to 0.5 TPa with some interstitial defect. Researchers worked on the mechanical properties of monolayer graphene using mathematical modelling and numerical simulations [10–12]. Young's modulus of nanometer thick suspended graphite sheet was experimentally investigated with force–displacement measurements by AFM on a strip of suspended graphene [13]. Circular membranes of few-layer graphene (FLG) have also been studied by force–volume measurements in AFM [14]. Recently, the fracture strength of isolated monolayer graphene has been studied by AFM. The fracture strength is 130 GPa [15].

8.2.3 Optical Properties

Black graphite becomes highly transparent when thinned down to a graphene monolayer. The electrons absorb light in the ultraviolet (UV) region but not in the visible region. So, pure monolayer graphene appears transparent to the human eye. This optical characteristic combined with the excellent conductivity of graphene-based materials will be considered as a replacement of silicon in the near future. This one-atom thick crystal can be seen with the naked eye because it absorbs approximately 2.3% of white light.

Indeed, in the visible range, thin graphene films have a transparency that decreases linearly with the film thickness. For 2 nm thick films, the transmittance is higher than 95% and remains above 70% for 10 nm thick films [17]. Bae et al. [18] have shown using numerical formula that absorption of light is directly proportional to the number of layers. Each layer has an absorbance value equal to 2.3%. Graphene can also become photoluminescent by manipulating the band gap energy. There are two methods of incorporating the photoluminescent property in graphene. The first method applies cutting down graphene in nanoribbons and quantum dots, while the second method deals with the chemical treatment with different gases, to pamper the connectivity of the π orbital electrons.

For example, Gokus et al. [19] introduced the luminescent property in graphene by oxygen plasma treatment. This alters only the top layer to obtain hybrid structures, while keeping the seed layers unaffected.

The combination of high film conductivity, optical transparency, chemical and mechanical stability immediately suggests employing graphene as transparent electrode for solar cells or liquid crystal and also as processable transparent flexible electrode material.

8.2.4 Electronic Transport

Suspended monolayer graphene has an amazingly high electron mobility at room temperature, which is 20,000 $cm^2/V/s$ [20–22] reported till date. As demonstrated by Stander et al. [23], the mobilities of holes and electrons are nearly the same. The mobility is independent of temperature between 10 and 100 K [10–12], which implies that the dominant scattering mechanism is defect scattering. The mobility is proportional to the carrier relaxation time and inversely proportional to the carrier effective mass. The electron mobility is often greater than hole mobility because quite often the electron effective mass is smaller than hole effective mass. The relaxation times are often of the same order of magnitude for electrons and holes, and therefore, they do not make too much difference. In order to increase the speed of a device, one has to choose materials with small electron and hole effective masses and long relaxation times, i.e. where the electrons and holes do not have to experience too much collision on crystal imperfections, impurities, etc. The scattering of charge carriers on defects in a crystal can significantly reduce the mean free path and the elastic scattering time which also determines mobility; thus, mobility decreases. If the room temperature mobility is 200,000 $cm^2/V/s$ at a carrier density of 10^{12} cm^{-2}, the subsequent resistivity of the graphene sheet would be 10^{-6} $\Omega \cdot cm$ [14,15] which is less than the resistivity of silver, the well-known substance with lowest resistivity at room temperature. But for graphene on SiO_2 substrates, scattering of electrons by optical phonons of the substrate is a larger effect at room temperature than scattering by graphene's own phonons. This limits the mobility to 40,000 $cm^2/V/s$.

Doped graphene with various species (some acceptors, some donors) as reported by Schedin et al. found that the initial pure graphene structure can be recovered by gradual heating of the graphene sheet in vacuum. They also reported that there is no change in carrier mobility if the chemical dopant concentrations exceed 10^{12} cm^{-2} [24].

The observable fact seen in graphene is its minimum conductivity that is approximately $4e^2/h$ irrespective of the applied voltage, where e is the electron charge and h is Planck's constant [10]. The minimum conductivity in graphene is really an unexpected phenomenon according to Geim and MacDonald [11]. Dirac points signify that the concentration of carriers at that point is nearly zero, so conductivity is also zero but this will not happen probably due to the uniqueness of graphene's electrons. It behaves as a wave rather than a particle. With the electron behaving as a wave, conduction continues to occur even if the carrier density near the Dirac points is a zero. Ripples in the graphene sheet or ionized impurities in the SiO_2 substrate can cause the electric conduction [25].

8.2.5 Anomalous Quantum Hall Effect

The conventional Hall effect demonstrates that an electric current, flowing through a rectangular metallic conductor, under the presence of a transverse magnetic field, produces a potential difference (Hall potential) between the two opposite faces. The ratio of the potential difference with respect to the current which is called Hall resistivity is directly proportional to the applied magnetic field. Hall effect is generally employed for the measurement of this magnetic field. It was observed that in a 2D structure, Hall resistivity becomes quantized at a temperature close to the absolute zero, holding an approximate value of h/ne^2 (where h is the Planck's constant, n a positive integer number and e is the electric charge). Hall resistivity was found only for the odd value of n which is a consequence of quantum mechanical effect named Berry's phase. This phenomenon is called quantum Hall effect. Compared to other 2D electron system, graphene shows some peculiar nature in characterizing the quantum Hall effect. For SLG, however, the Hall conductivity is quantized if the filling factor is a half-integer whereas for conventional 2D systems, the Hall resistance emerges when the filling factor is an integer. The formation of discrete Landau levels is crucial to the quantization of the Hall resistance and conductivity.

This unusual event is due to the quasi-particle excitation in graphene, which can be explained by the massless Dirac's equation of the pseudo-relativistic theory. This will lead to the appearance of a Landau level with zero energy and hence a shift of ½ in the filling factor with respect to the conductivity quantization. This is probably due to the fact that, in graphene, the magnetic energy of electron is 1000 times higher than that of other materials.

Interestingly, electrons in graphene show 100% transmission rate through a potential barrier of any size. The quantum mechanical explanation for this observation is that the barrier can put an obstruction to the electrons travelling in a particular valley by changing its orientation, but it does not obey the rules of 'conservation of chirality'.

Geim et al. [26] and Kim et al. [27] reported the quantum Hall effect in graphene. It was a monolayer of carbon stacked on a thin silicon dioxide layer on top of a doped silicon substrate which served as a gate electrode. In his paper, Geim proved how gate voltage dopes graphene p-type (as shown by the positive Hall coefficient R_H, through zero, to n-type. It is a good evidence of how Hall coefficient R_H and electrical conductivity σ both extrapolate to zero when the Fermi level passes through the Dirac points.

8.2.6 Magnetic Properties

A very astonishing event found in graphene is the missing atoms in the hexagonal structure (called vacancies), which act as tiny magnets (they have a magnetic moment) and have a great influence on the electrons in graphene which carry electrical currents, indulging a considerable amount of extra electrical resistance at low temperature. This effect is known as the Kondo effect. Azyev et al. [28] have worked on it. It was also suggested that the 'zigzag' edges of the graphene nanostructure are also responsible for the magnetic effect [29]. Wang et al. [30] have also reported that the reduction of graphene oxide in hydrazine atmosphere showed the ferromagnetism effect. Graphene mainly shows inhomogeneous magnetism below 20 K as reported by Joly et al. [31], probably due to the presence of charged impurities in the SiO_2 substrate.

The whole things were discovered by researchers from the University of Maryland. They also worked upon the position of the vacancies and found that if the vacancies are arranged in a right order ferromagnetism would be observed. Adsorption of water molecule/acid molecule and intercalation with potassium nanocluster can cause decrease in the magnetic effect as reported elsewhere [32]. This powerful tool can be used in nanoscale sensors of magnetic fields and could also be useful in spintronics. Intrusion of foreign molecules through physisorption in the graphitic layer mechanically reduces the intergraphene layer distance which disturbs the alignment of magnetic moment and the net magnetic moment thus reduced [33].

8.2.7 Thermal Properties

There is a very low carrier density in non-doped graphene; the electronic contribution to thermal conductivity is negligible. The thermal conductivity of graphene depends on the phonon transport phenomenon. At high temperature, conduction occurs through diffusion, and at low temperature, ballistic conduction occurs as there is no scattering of electrons. Due to the absence of impurities, defects in a pure graphene and the mean free path are longer than other metals [34]. At room temperature thermal conductivity of a suspended monolayer graphene is 6000 W/m/K, which is much higher than graphite [35]. Zhu et al. [36] have reported a thermal conductivity value of about 5000 W/m/K for a suspended monolayer graphene, produced by mechanical exfoliation. A trench was used to suspend the graphene. A laser beam was focused on the centre of the suspended graphene. The heat flowed outward from the centre of the graphene to the extreme end. The heat loss via air was found to be less than through the graphene [37]. The temperature rise in the heated graphene causes a red shift of the Raman G peak because of bond softening. Another study exhibits a thermal conductivity of 2500 W/m/K (at 350 K) obtained from CVD-grown graphene onto the thin silicon nitride membrane; the silicon nitride was coated with a thin layer of gold for better thermal contact [38]. It has been recently reported [39] that the thermal conductivity of micromechanically exfoliated graphene deposited on a SiO_2 substrate is about 600 W/m/K, which is much more than Cu (385 W/m/K).

8.3 Characterization Techniques

Before detailing the graphene production methods, it is appropriate to provide the reader with the tools to locate, recognize and characterize graphene. Rather than a complete list and description of relevant techniques, a basic demonstration is summarized here so that any researcher working on graphene should get access to in order to facilitate graphene-based projects. The characterization techniques used are (1) optical, (2) scanning probe and (3) electron microscopy as well as (4) GAXRD (glancing angle x-ray diffraction study), (5) Raman spectroscopy techniques, etc.

8.3.1 Optical Microscopy

The use of a suitable substrate while fabricating a graphene sheet is an important issue as the substrate can maximize the optical contrast of the monolayer graphene sheet in the wavelength of utmost sensitivity for the researchers. The easiest way to differentiate between different thicknesses of graphene is optical microscopy while using Si as a substrate material. For SiO_2/Si substrate, the deposited graphene adds a small optical path to the Fabry–Perot cavity up to 0.3 μm thick oxide dielectric where the maximal contrast occurs at a wavelength of 550 nm, visualizing the green channel image normalized by its value on the bare substrate. A more difficult study has been performed where the whole spectrum of the white light has been taken in consideration and fully explained by Berger et al. [40]. Park et al. [41] studied the optical micrograph of FLG prepared by CVD. FLG shows a monotonic increase in absorbance as the photon energy increases up to 4.5 eV where an absorbance maximum appears.

8.3.2 Field Emission Scanning Electron Microscopy

For effective picturization of graphene layer, field emission scanning electron microscopy (FESEM) needs to meet these requirements. As graphene is basically a nanostructured material, to study the morphology of graphene, a small beam spot size corresponding to a high spatial resolution is highly required. The graphene monolayer seems to be transparent to high-energy electron beams. Since the ideal imaging information for FESEM is the secondary electrons generated by the primary beam in the sample, low beam energy is required to image this ultra-thin material. Moreover, it restricts the sample from damage due to beam energy. The imaging of graphene film is always preferred on insulating substrates. Further, many features in graphene are difficult to image because of poor contrasts. A FESEM with contrast control facility is perfect for graphene monolayer as this is very much levelled and distinct making it challenging for characterization. Figure 8.7 shows one of the FESEM results on Ni particles.

If the low-energy electrons can be detected to give the best surface contrast and topographic sensitivity, then the ideal technique would be low-voltage FESEM which is ideal for graphene imaging.

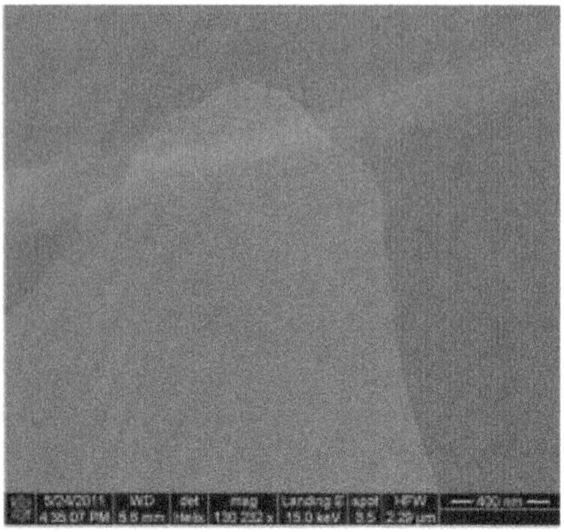

FIGURE 8.7
FESEM image of flat graphene film. (From Chuan, X.-Y., *Int. Nano Lett.*, 3, 1, 2013.)

8.3.3 Atomic Force Microscopy

The first optical resolution technique used to identify the thinnest graphitic flakes in front of the world was AFM and that was indeed one-atom thick monolayer, i.e. graphene. An SLG on oxidized wafers consistently appears to be 0.8–1.2 nm thick but surprisingly an additional layer was found on top of it totalling the expected 0.35 nm thickness [43,44]. The origin of this extra layer is still invisible since pure van der Waals interaction between silica and graphene has no credit for it. But it can be concluded that probably the ambient while in deposition is responsible for this topping. When a portion of SLG is managed by an AFM tip, it reveals a trace that is just the same shape and an approximate height of 0.3–0.6 nm depending on the AFM surrounding atmosphere that consists of atmospheric hydrocarbons between the graphene and SiO_2 surfaces by capillary condensation. AFM soft imaging modes are the best way to examine the topological quality of substrate-supported graphene samples while fabricating devices. Figure 8.8 shows one of the AFM results found in graphene transferred on SiC substrate.

8.3.4 Diffraction Imaging Electron Microscopy

There are no distinct grain boundaries observed in graphene. Working on graphene grain boundaries, researchers from Cornell University concluded that graphene doesn't grow in perfect sheets – it rather develops in pieces that resemble patchwork quilts (Figure 8.9). The meeting point of those patches is called grain boundaries, and the researchers are studying those boundaries.

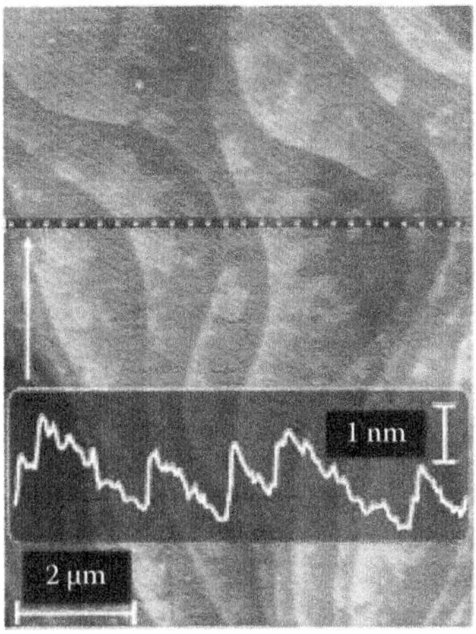

FIGURE 8.8
AFM image of graphene grown on SiC substrate. (From Szroeder, P. et al., *J. Solid State Electrochem.*, 18, 2555, 2014.)

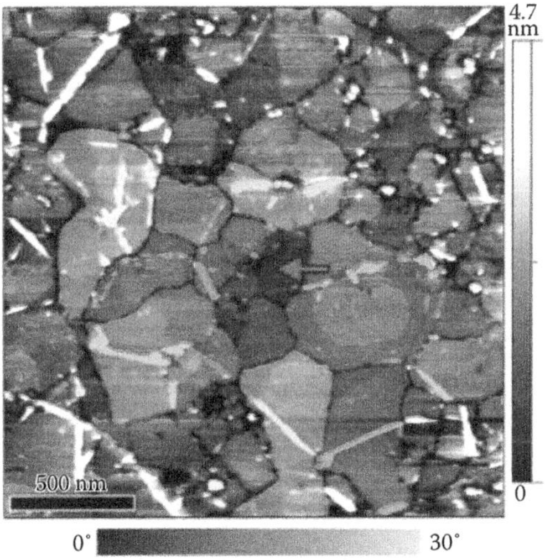

FIGURE 8.9
Graphene grain boundaries. (From Biro, L.P. and Lambin, P., *New J. Phys.*, 15, 035024, 2013.)

The researchers grew graphene on copper and then etched away the Cu by a novel way to peel them off as free-standing, atomic thin films. The appearance of graphene was pictured using diffraction imaging electron microscopy and used a colour to represent the angle that electrons bounced off at. The use of different colours shaped an easy way to visualize the graphene grain boundaries.

8.3.5 Transmission Electron Microscopy

In a transmission electron microscopy (TEM), a broad beam passes through a thin sample producing an image due to diffraction or mass thickness contrast. This image is subsequently magnified by a set of electromagnetic lenses and recorded on a photographic film. Electron diffraction studies can be used to qualitatively distinguish a single from a bilayer since both exhibit a sixfold symmetry [47] for the two objects as predicted by Horiuchi et al. [48].

TEM has been exploited to understand the structure of graphene. Only TEM of cross-sectioned graphite films would allow for direct measurement of the film thickness, quality and uniformity.

TEM analysis is a good characterization tool for the analysis of electronics and crystal structure of graphene. Robert Colby [49] analyzed in the literature that cross-sectional TEM would provide direct, straightforward analysis of film thickness and quality, as well as provide a means to investigate specific defect structures (e.g. wrinkles). Heer et al. [50] reported the TEM image of epitaxially grown graphite on SiC substrate.

The atomic resolution image of SLG (shown in Figure 8.10) shows a perfect single-crystal structure in the displayed region, where no defect can be seen.

8.3.6 Raman Scattering

The thickness of graphene sheet can be obtained by the electron–phonon interactions in Raman spectra. The number of layers in a graphene film can be calculated by measuring the intensity, profile and location of the G and 2D bands. If the number of layers increases, 2D band changes its shape and position (Figure 8.11) while G-band peak position shows a downshift with number of layers (Figure 8.11). The Raman spectra of graphene comprise the

FIGURE 8.10
TEM images of reduced graphene oxide. (From Firdhouse, M.J., Lalitha, P., *Int. Nano Lett.*, 4, 103, 2014.)

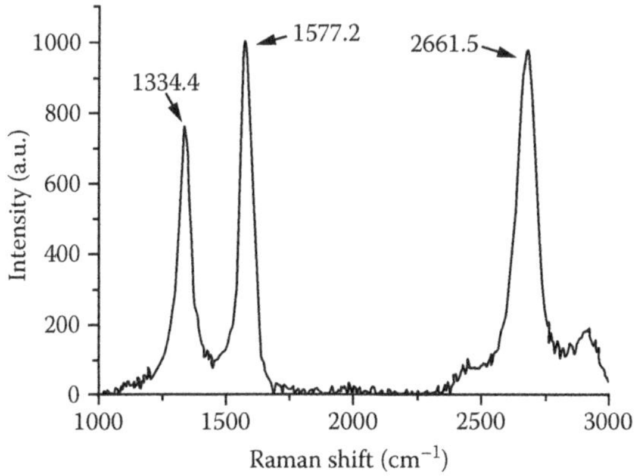

FIGURE 8.11
A typical Raman spectrum of few-layered graphene obtained by the arc-discharge process. The three intense features are the D band at 1334 cm^{-1}, the G band at 1577 cm^{-1} and the 2D band at 2662 cm^{-1}. An in-plane crystallite size La = 5.9 nm was calculated using the equation La (nm) = 4.4 (IG/ID). The intensity ratio $IG/I2D$ of the product is ~1.03, which is comparable to the value for three layers of CVD-grown graphene ($IG/I2D$ ~1.3). (From Wu, Y. et al., *Nano Res.*, 3, 661, 2010.)

G peak placed at 1580 cm^{-1} and 2D peak at ~2700 cm^{-1}. The G peak at around 1560 cm^{-1} corresponds to the $E_{2}g$ phonon at the centre of the Brillouin zone. The D peak, at 1360 cm^{-1}, is due to the out-of-plane breathing mode of the sp^2 atoms and is active in the presence of a defect [51]. The D band is an efficient probe to assess the level of defects and impurities in graphene.

8.4 Synthesis of Single-Layer Graphene/Few-Layer Graphene

The synthesis of monolayer graphite was invented at early 1970s, when Lang et al. [52] tried to form the mono- and multilayered graphite by thermal decomposition of carbon on single-crystal Pt substrates. But due to the lack of reliability of the properties of such sheets, some extended studies have been made to produce graphene. Novoselov et al. [4] have been awarded Nobel Prize for the discovery of graphene in 2004. They first proved the repeatable synthesis of graphene through exfoliation. Many other production methods have been reported for a single layer of graphene, for example micromechanical cleavage of graphite, CVD, epitaxial growth and oxidation of graphite which is beyond our scope of limit.

8.4.1 Micromechanical Exfoliation

Graphite is a polymorph of the elemental carbon. It can also be stated as a stacked layer of many graphene sheets, binding tightly by weak van der

Waals force. It is easy to realize that it is possible to produce graphene from a high-purity graphite sheet, if these bonds are broken. As graphene is basically a 2D lattice, due to the large lattice spacing in *c* direction and also weak π bonding in the *c* direction, it is a trouble-free attempt to obtain graphene sheets through the exfoliation of graphite. Experimentally, exfoliation of graphite has been done in many ways including chemical, mechanical and thermal methods [53]. The first step was made by Viculis et al. [57], who have used potassium metal to intercalate a pure graphite sheet and then exfoliate it with ethanol to form thinner sheet of carbon atoms. TEM analysis showed the presence of 40 ± 15 graphene layers in each sheet. Novoselov et al. [55] reported FLG and even SLG structure by exfoliation technique. The scotch tape 'synthesis' method is very basic, but it can't be adjusted in the device formation and not able to produce large quantities of graphene. Mechanical exfoliation technique for producing graphene was first tested on the oxidized silicon substrate. Researchers at Georgia Institute of Technology have used silicon carbide as the substrate material and produced graphene by the thermal decomposition of SiC [42]. At high temperatures, the silicon vaporizes leaving behind the hexagonal carbon lattice. In this process, a carbon layer acts as a seed layer on top of which graphene layer is formed.

Chang et al. [56] equipped graphene on a fibre ferrule. They have used scotch tape exfoliation method for highly ordered pyrolytic graphite (HOPG) to make thin graphitic layer, then made a contact between cleaned fibre ferrule with the graphitic layer. They have used scotch tape exfoliation method for highly ordered pyrolytic graphite (HOPG) to make thin graphitic layer then made a contact between cleaned fibre ferrule with the graphitic layer following the subsequent separation of ferrule from the graphite.

Wong [57] fabricated nanomechanical graphene drum structures for resonant mass sensing applications with 10^{-20} g/Hz sensitivity, which vibrate at a frequency greater than 25 MHz using mechanical exfoliation of graphite onto pre-etched circular trenches in silicon dioxide on a silicon substrate.

Tang et al. [58] have exfoliated high-quality graphene flakes from graphite by solvothermal treatment. Due to the presence of oil/water interface, graphene can be easily and quickly estranged from the graphene/NMP solution. Shukla et al. [59] produced mm-sized single to FLG by bonding bulk graphite to borosilicate glass after subsequent exfoliation, to leave single or few layers of graphene on the substrate.

Hiura et al. [60] and Ebbesen and Hiura [61] observed folding and tearing of graphitic sheets that formed spontaneously during scanning due to the friction between the tip and HOPG surface. It was found that the folding and tearing of graphitic sheets follow well-defined patterns due to the formation of sp^3-like line defects in the sp^2 graphitic network, occurring preferentially along the symmetry axes of graphite. The curved portion is accompanied with ripples, in order to release the strain and stabilize the electronic structure in the bent region. The possibility of creating various types of 3D graphene structures through folding and refolding of graphene sheets in different ways has been discussed. Instead of forming graphene sheets spontaneously during tip scanning on HOPG, Roy et al. [62,63] have tried to fold and unfold the graphene sheets in a more controllable way through modulating the distance or bias voltage. In all these experiments, one must first locate step edges using AFM over a large sample surface area, and thus, the entire process is not well controlled. Furthermore, it is also difficult to obtain large-size graphene sheets using this method.

Ever since this work, mechanical exfoliation has become the method of choice for producing graphene with highest quality. Many variations in original exfoliation techniques have been developed and applied to different types of graphites. Although the mechanical exfoliation technique has been improved significantly, its primary drawbacks still remain. Its low productivity does not allow synthesis of graphene in large quantities. It is also incompatible with standard Si processes. The former might be overcome by chemical exfoliation and CVD, while the latter may be avoided by using epitaxial growth.

8.4.2 Chemical Exfoliation

Mechanical exfoliation suffers from one major disadvantage; this process results in the formation of structural defects which is probably the reason of low conductivity due to scattering of electrons in the graphene layer. Fabrication of graphene sheets via chemical routes is high area of interest nowadays. An understanding with different chemical processes like intercalation, oxidation, exfoliation, reduction, fictionalization and dispersion should be developed first and should be familiar with the reaction processes.

A significant achievement has been made recently showing that graphite could be exfoliated using liquid chemicals to give defect-free monolayer graphene [65,66]. One of the popular methods to form graphenes through chemical exfoliation is immersing the graphite in a mixture of sulphuric and nitric acid [67,68] for a specific period of time. The acid molecules enter into the graphitic layers to increase the interlayer spacing, forming alternating layers of graphite and intercalant. The graphitic layer gradually etched out, and the thickness of the graphite decreases. After the formation of the thin graphite sheets, the rapid evaporation of the intercalants at elevated temperature produces high-yield graphene layer with high scalability. The chemical exfoliation is generally accomplished in two process steps.

The graphene oxides (GOs) can also be used for the formation of graphene sheets. Thicker GO consequently exfoliated to form thin GO sheets using different techniques [69]. A chemical, electrochemical or thermal reduction process is then performed to convert the GOs into graphene sheets. Viculis et al. [54] reported the synthesis of graphite nanoplatelets using volatile organic compound (ethanol) as the exfoliating agent, which thickens down to 2–10 nm by using acid (potassium or cesium at 200°C)-intercalated graphite with microwave heating as the last step.

Hernandez et al. [47] worked on surfactant–water solutions with the aid of ultrasound to disperse graphite, the outcome of which was large-scale exfoliation to give large quantities of few-layer graphene. Lotya et al. [70] reported single- to few-layer graphene preparation, by making dispersion of graphite powder in sodium dodecylbenzenesulfonate (SDBS), then sonication, to transfer the graphite into graphene. Similar attempts were found to produce graphene either from graphite or from graphite oxide powder and using special solvents [71–74].

8.4.3 Epitaxial Growth on Silicon Carbide (SiC)

Epitaxy means the growth of a single-crystal film on top of a crystalline substrate. Epitaxially grown layers are purer than the substrate. Epitaxial graphene (EG) is basically a single layer of graphite, a hexagonal array of carbon atoms extending over two dimensions endlessly. EG on silicon carbide has played a pivotal role in this development: it was the first to be proposed as a platform for graphene-based electronics; the first measurements on graphene monolayers were made on EG, and the graphene-electronic band structure was first measured on EG. The EG program, initiated in 2001 at the Georgia Institute of Technology, has spearheaded graphene-based electronics and developed methods to produce electronic-grade EG. It is important to note that within the next 10 years, fundamental property limitations of silicon will inhibit the ability to fabricate operational devices and circuits due to continuing device size reduction. EG has extraordinary electronic properties that offer the possibility of greatly enhanced speed and performance relative to silicon; this material may serve as the successor to silicon in integrated circuits and microelectronic devices. This approach offers the advantage that high-quality layers can be grown on large-area substrates. Most of the graphene researchers concentrated on exfoliated EG, where EG flakes are obtained by peeling layers from graphite.

A major step towards the wafer-scale applications of EG has been the graphitization of Si-terminated SiC(0001) in inert gas atmosphere. Hu et al. [75] fabricated orientation-controlled epitaxial SLG over the Cu(111) film on sapphire, while a polycrystalline Cu film deposited on a Si wafer gives non-uniform graphene with multilayer flakes. Carbon isotope–labelling experiment indicates rapid exchange of the surface-adsorbed and gas-supplied carbon atoms at the higher temperature, resulting in the highly crystallized graphene with energetically most stable orientation consistent with the underlying Cu(111) lattice. Hwang et al. [76] have presented van der Waals epitaxial growth of graphene on sapphire by CVD without a metal catalyst. Grandthyll et al. [77] have worked on the epitaxial growth of graphene on transition metal surfaces by ex situ deposition of liquid precursors (liquid-phase deposition [LPD]) is compared to the standard method of CVD. The performance of LPD strongly depends on the particular transition metal surface.

Hirokazu Fukidome et al. [78] experimented the growth of graphene on SiC substrates by thermal decomposition of surface layers. This graphene-on-SiC technology is the abdication of the well-established Si technologies and the high production cost of the SiC bulk crystals. The substrates are p type (1–10 Ω cm). The epitaxy of GOS consists of two steps. The first step is the epitaxial formation of 3C-SiC thin films on silicon substrates by gas-source molecular-beam epitaxy using monomethylsilane. The second step is graphitization of the SiC thin films by annealing the films in vacuum with a resistive heating at 1250°C for 30 min. Vacuum results in the sublimation of the silicon atoms while the carbon-enriched surface undergoes restructuring to produce graphene layers [79–83].

One of the highly trendy techniques of graphene growth is thermal decomposition of Si on the (0001) surface plane of single crystal of 6H-SiC [84]. Graphene sheets were found to be formed when H_2-etched surface of 6H-SiC was heated to temperatures of 1250°C–1450°C, for a short time (1–20 min). Graphene, epitaxially grown on this surface, typically had 1–3 graphene layers. A comparable process, as reported by Rollings et al., have produced graphene films, as low as one-atom thickness [85].

Hass et al. [5] have presented a comprehensive review on this topic, covering the issues of graphene growth on different faces of SiC and their electronic properties. The process of growing graphene on SiC seems lucrative, particularly for the semiconductor industries. However, issues like controlling thickness of the graphene layers and repeated production of large-area graphene have to be solved, before the process can be adopted at an industrial scale. This process still suffers from surface roughening by the formation of deep pits and the limited lateral extension of crystallites [86].

8.4.4 Chemical Vapour Deposition

A highly pure large-quantity graphene can be procured using CVD technique. For graphene preparation, two types of CVD processes are engaged. First one is low-pressure CVD, wherein the advantage is production of more uniform graphene layer by prevention of many unwanted reaction due to low-pressure. In thermal CVD process, the evaporation of metallic layer due to the slow increase in temperature in the CVD reactor ultimately affects the growth of graphene. The synthesis of graphene by using plasma-enhanced chemical vapour deposition (PECVD) technique can circumvent this problem by lowering down the process time with the decrease in substrate temperature. CVD graphene is created in two steps, the precursor, pyrolysis of a material to form carbon and the formation of the carbon structure of graphene from the dissociated carbon atoms. For the pyrolysis of precursor, a high temperature is required to generate huge amount of heat. So, a catalyst is essential to reduce the temperature of reaction. CVD graphene is transferable to different substrates by increasing the adaptability of the device for various applications.

Graphene was found to be formed on the crystal surface, by thermal decomposition of either ethylene, acetylene or methane (preadsorbed on crystal surface at room temperature) at 1000 K. Sutter et al. [87] prepared graphene by thermal decomposition of preadsorbed C atoms on Ru(0001) surface, leading to the formation of macroscopic (more than 200 μm) single-crystalline domains of single- to few-layer graphene. Using a similar kind of approach, graphene monolayers were grown on single-crystal Ru(0001) surface at ultra-high vacuum condition (4×10^{-11} Torr) [88]. Before the synthesis process, Ru crystal was cleaned by repeated cycles of Ar+ sputtering/ annealing and exposure to oxygen and heating to high temperature. There are several reports for graphene synthesis on other transition metals also including Ir, Ni, Co and Pt facilitating them for wide usage in sensor applications.

To avoid the high cost of single crystal SiC, FLG (few layer graphene) films were grown on polycrystalline Ni foils by CVD of methane under atmospheric pressure in a decisive manner. The cooling rate of the metal substrates is very crucial, as carbon can be precipitated during cool down [89]. It is beyond the scope of discussion on different transition metals; here, the discussion will be limited to Ni, whose surface chemistry with hydrocarbons has been investigated very accurately [90–93]. In this work, a biodegradable, low-cost precursor, camphor was used to synthesize graphene on Ni foils. Camphor was first evaporated at 180°C and then pyrolyzed, in another chamber of the CVD furnace, at 700°C–850°C, using argon as the carrier gas. Graphene sheets were observed on the Ni foils after natural cooling having approximately 35 layers found in high-resolution transmission electron microscopy images. This study opened up a new horizon for graphene synthesis, though several issues like controlling the number of layers and minimizing the folds were yet to be solved.

The conductance of graphene monolayers which can be controlled by a gate electrode has inspired many of the researchers to continue their research on both the fundamental physics and the technological domain [94–96]. Pollard et al. [97] have synthesized the graphene films by annealing Ni films deposited on SiO_2/Si at 800°C under vacuum conditions ($\sim 5 \times 10^{-10}$ mbar). The film was stored under atmospheric conditions for typical periods of a few days before placing it in another chamber. Following the growth, Ni is removed by etching, and the graphene is transferred as a single continuous layer onto a separate substrate. The fraction of monolayer graphene is investigated using optical and electron microscopy and Raman spectroscopy and is shown to be >75%.

At the same time, Li et al. [98] proved that at high temperatures, the isotopic carbon diffuses into the Ni first, mixes and then segregates and precipitates on the surface of Ni forming graphene that was determined by the peak position of the Raman G-band peak. Shelton et al. [99] theoretically explained that this segregation refers to compositional heterogeneity in thermal equilibrium which corresponds to a 'one-phase' field, while precipitation refers to inhomogeneities that arise as a result of equilibrium 'phase separation'.

It has been proposed that the CVD growth of graphene on Ni occurs by a C segregation or precipitation process whereas graphene on Cu grows by a surface adsorption process [100] where the spatial distribution of 12C and 13C follows the precursor time sequence and the linear growth rate ranges from about 1 to as high as 6 μm/min depending upon Cu grain orientation.

The use of Ni thin film rather than Ni foil is to minimize the saturation time since the solubility of C in Ni is high, about 0.9 at % at 900°C. It was found that the graphene growth on Cu substrate was limited by its internal characteristics, probably due to limited solubility of C in Cu.

The process was claimed to be a surface-catalyzed process rather than a precipitation process, as has been reported for Ni [93]. In this work, both normal methane and $13CH_4$ (99.95% pure) were introduced to the growth chamber in a specific sequence. Ni film is transferred onto a SiO_2/Si wafer by poly(methyl methacrylate) (PMMA) similar to the reported method [98,100].

The efficient transfer approach is the use of PMMA, which was spin coated on top of the synthesized graphene films lying on Si/SiO_2/Ni substrates. The wafer was dipped into a nickel etchant solution at 90°C for 2 h to etch away the nickel film and leave a free-standing PMMA film with the deposited graphene attached to it. PMMA/graphene film was transferred to other substrates (Si/SiO_2, glass, etc.), and then acetone was used to dissolve the PMMA residues and leave clean graphene films on the target substrate surface.

Rybin et al. [102] have optimized experimental conditions to prepare actual monolayer graphene. To start with, the nickel foil was annealed in hydrogen at a pressure of 0.5 bar at 1000°C for 10 min. Then methane with 99.99% purity was introduced to reach the concentration ranging from 5% to 50% for 2 min. After etching of the Ni foil with $FeCl_3$, the graphene film was transferred to other substrate.

The large-scale preparation of graphene was first made by Lewis Gomez De Arco et al. [103] by the same CVD technique over the complete wafers. They have followed the same CVD technique by heating the substrates under a flow of 600 sccm of H_2 up to 800°C and simultaneously methane gas at a flow rate of 100 sccm over the substrate for 2 min. They have proved that diluted methane was key to the growth of SLG and FLG (less than five layers). Addition of hydrogen improved the quality of graphene. The possible c for this is that hydrogen is known to selectively etch amorphous carbon defects that can serve as secondary nuclei for competing film growth [104]. This may be the probable cause for the H_2 use. They have immersed the graphene-on-nickel sample into a nickel etchant solution. This process removed nickel and left graphene films deposited on the underlying Si/SiO_2 substrate.

Park et al. [41] have reported the similar CVD approach with nickel thin film (300 nm). The Ni-coated substrate was positioned at the centre of a quartz tube and heated to 1000°C at a 40°C/min heating rate, under a flow of argon and hydrogen (Ar/H_2 = 1, 1000 sccm). The substrate was kept at 1000°C in the Ar/H_2 flow for 20 min to anneal the Ni film. The CVD growth of FLG was conducted at 960°C–970°C, under a continuous flow of carrier gases,

Ar and H_2, mixed with less than 5 vol.% methane ($CH_4/Ar/H_2 = 250:1000:4000$ sccm). After the CVD reaction, the sample in the quartz tube was cooled down to 400°C at the rate of 8.5°C/min, under a flow of argon and hydrogen. They have also measured the sheet conductivity as a function of temperature from 300 to 4 K by four-probe measurement method. The sample was placed into the sample chamber and annealed at 120°C in vacuum (10^{-7} mbar) for 24 h before the measurement.

Atomic hydrogen then acts as an etching agent reducing preferably amorphous carbon that contains unsaturated dangling bounds. For CVD graphene growth, we found that a critical amount of hydrogen is also necessary to synthesize FLG because hydrogen keeps a balance between the production of reactive hydrocarbon radicals and the etching of the graphite layer during the CVD process. If the ratio of methane to hydrogen is too low, the etching reaction becomes much faster than the formation of graphene layers. This was also experimentally proved in a recent work of Kong's group [105]. $FeCl_3$ solution etches the nickel layers in a mild pH value, without causing the formation of gaseous products or precipitates. Gaseous products, as is produced by HNO_3 etching, damage the graphene structure [106]. The presence of amorphous carbon at the bottom side of the FLG strongly supports a growth mechanism of graphite on nickel by CVD as suggested by Obraztsov et al. [107].

Zheng et al. [45] demonstrated metal-catalyzed crystallization of amorphous carbon to graphene by thermal annealing. The thickness of the precipitated graphene was controlled by the thickness of the initial amorphous carbon layer. This is in contrast to CVD processes, where the carbon source is virtually unlimited and controlling the number of graphene layers depends on the tight control over a number of deposition parameters. Based on the Raman analysis, the quality of graphene is comparable to other synthesis methods.

Chen et al. [42] described a method for the bulk growth of mono- to few-layer graphene on nickel particles by CVD from methane at atmospheric pressure. A graphene yield of about 2.5% of the weight of nickel particles used was achieved in a growth time of 5 min. After the growth of graphene, the nickel particles can be effectively removed by a modest $FeCl_3$/HCl etching treatment without degradation of the quality of the graphene sheets.

Kondo et al. [108] have first described the low-temperature synthesis of graphene using acetylene gas on nickel films starting at 650°C as the barrier energy for breaking the C–C bond is lower than for C–H bonds and fabrication of top-gated field-effect transistors (FET) without using transfer processes. Typical flow rates were as follows: acetylene was 12 sccm and hydrogen was 24 sccm, with a total pressure of 800 mTorr. The decomposed carbon atoms then dissolve into the bulk of the Ni film as interstitials, forming solid solutions. It is during the cooling cycle that the segregation of the carbon atoms leads to the formation of the graphene layer(s); they also demonstrated that with Ni films thinner than 200 nm, the annealing step resulted in decomposition of the metal film into particulates, resulting in discontinuous layers.

Coraux et al. [109], Marchini et al. [110] and many other authors have utilized transition metal surfaces other than nickel (Pt, Ru, etc.) for the growth of graphene using either methane or ethylene gases as precursors at deposition temperatures of about 1000°C. Nandamuri et al. [111] have also reported reduced pressure CVD method. According to their scheme, single and multiple layers of graphene films were produced on polycrystalline nickel films using remote plasma-assisted CVD. Remote plasma was employed to eliminate the effect of the plasma electrical field on the orientation of the grown graphene films, as well as to reduce the growth temperature compared to conventional CVD. Regan et al. [112] given another example of using $FeCl_3$/HCl solution for etch back technique.

Zhu et al. [104] have proposed a growth mechanism for the graphene in PECVD chamber by plasma electric field direction. In this way, 71 graphene sheets are synthesized by a balance between deposition through surface diffusion of C-bearing growth species from precursor gas and etching caused by atomic hydrogen.

It need not be mentioned that the CVD process for making defect-free graphene layer is an unparallel approach in the field of nanotechnology. But the transfer of graphene to other substrate may not be always appropriate for large-scale preparation. To overcome this problem, direct growth of graphene on metal oxide surfaces is a promising effort that has been made by many researchers [117].

Very recently, Chang Mook Lee and Jaewu Choi [26] fabricated nanographene with an in-plane crystal size of 15 nm which was directly deposited on glass at 750°C by thermal chemical deposition, without using any additional transition metal catalyst. An additional layer of nickel was deposited on top of the graphene layer to reduce the annealing temperature (300°C). This study clearly shows that nanographene can be directly deposited onto glass and the crystallite size is controllable.

Epitaxial growth of graphene layers requires high temperatures and an ultra-high vacuum environment [113], which is expensive and thus may limit the widespread application of graphene. Ambient-pressure CVD has recently been reported to produce large-area films of 1–12 graphene layers [100], but a purification process is needed to eradicate the catalyst Ni particles in order to obtain clean graphene sheets. There are also many other synthesis method involved in graphene though it is not possible to explain all in the same chapter.

8.5 Graphene Oxide

Graphene oxide formerly called graphitic oxide is a compound of carbon, oxygen and hydrogen combined in a fixed ratio and is a very remarkable substance because its mechanical, electronic and optical properties can be controlled by applying thermal or chemical treatments to alter its structure. Graphene oxide is not just an arrangement of simple C and O, there are a lot of different function groups involved in GO. There are carbonyls, oxidative rings, alcohols, epoxides and other oxygen-based functional groups. In the last few years, it was studied that most oxidation occurs at the edges.

A novel graphene oxide paper-like material possessing a unique layered structure has been developed by directed-flow assembly method in which each graphene oxide sheets are interlocked mutually in a near-parallel fashion. The average modulus of graphene oxide paper was found to be 32 GPa [114]. The conventional Hummers' method developed in 1957 using $KMnO_4$ has been modified by the pre-exfoliation of the graphite by microwave heating. Vacuum filtration of colloidal dispersions of graphene oxide sheets through an Anodisc membrane filter yielded free-standing graphene oxide paper with thicknesses ranging from 1 to 30 mm after drying. The average elastic modulus and the fracture strength were found to be 32 GPa and 120 MPa, respectively. The mechanical properties of this 'graphene oxide paper' have been improved by chemical cross-linking between individual platelets by the use of divalent ions [115] and polyallylamine [116]. A self-assembled graphene oxide paper was also made at a liquid/air interface by evaporating the hydrosol of graphene oxide [117]. After annealing, its stiffness and tensile strength increased to such a value which is not reported till date. Graphene oxide prepared by a 'bottom-up' approach (Tang–Lau method) is much more affable compared to traditional 'top-down' method, in which strong oxidizers are involved. In 'bottom-up' approach, the precursor is glucose; the advantage of this process is that the thickness can be controlled more easily. Graphene oxide reduction with hydrazine has been reported [118] earlier.

A method chemically converted graphene that is superior than conventional Hummers' method provides a greater amount of hydrophilic-oxidized graphene using excess $KMnO_4$. The GO produced by this method is more oxidized than earlier and does not generate toxic gases.

Shen et al. [119] have reported an inexpensive, massively scalable, fast and facile method for the preparation of graphene oxide nanoplatelets. The basic strategy involved the preparation of graphite oxide from graphite through reaction with benzoyl peroxide (BPO), complete exfoliation of GO into graphene oxide sheets, followed by their in situ reduction to reduced graphene oxide nanoplatelets. The mechanism of graphene oxide producing is mainly the generation of oxygen-containing groups on graphene sheets. The synthesis of graphene oxide (GO) sheets on a large scale was developed by chemical exfoliation technique by simply controlling the oxidation and exfoliation procedure as reported by Zhang et al. [120]. GO samples were prepared under different conditions, which all have excellent water dispersion. It is found that when longer oxidation times and more oxidants are used, the mean size of the GO sheets, the Gaussian distribution, decreases.

A cost-effective method for mass production of graphene-based devices is to first produce chemically modified graphene, it starts with graphene oxide (GO), and then reduces it to obtain graphene for device applications. Large-quantity GO can be easily produced by the chemical exfoliation of graphite through oxidation and the subsequent dispersion in water [121]. The basal plane and edges of GO are decorated with oxygen functional groups [117,121], making GO highly soluble in water. Single-layer GO sheets can be generated by simple sonication of hydrophilic graphite oxide in water. GO is electrically insulating, owing to the disruption of the sp^2-bonded graphitic structure by the attachment of electronegative oxygen atoms [122].

The automatic ripple formation in suspended graphene [123], where the ripples have rare sinusoidal profile and whose presence can be controlled by temperature applying the negative thermal expansion of graphene, might also be a milestone to be achieved for future researchers.

8.6 Potential Application

Mono- and multilayer graphene have achieved the milestone of present era and have revealed limitless scientific and technological breakthrough with nanodevice applications, particularly in optoelectronics and sensor applications. Although the nanocrystalline semiconducting metal oxide–based gas sensors have significant usage in different parts of life, the most excellent sensor would be the one that can identify atomic level presence of the toxic gases. In view of that, graphene has opened a new horizon that assures an ultra-sensitive and ultra-fast electronic sensor. The applications of this wondering material are as follows: battery, FET, integrated circuit, transparent conducting electrode, solar cell, etc. Our area of interest is graphene-based chemical sensors.

8.6.1 Graphene Sensors

Sensor could be one which is able to determine a number of environmental pollutants including toxic and hazardous gases. Graphene as a super material for its unique electronic properties showed very high electrical conductivity and very low electronic noise. One of the major advantages of the graphene-based nanosensor is that it can be made ultra-sensitive and can be utilized for repetitive measurements. SLG has a very large surface area (2600 m^2/g).

In traditional room temperature metal oxide thin film–based sensors, the adsorbed molecules are not completely detached from the sensing layer surface, leaving behind some residue on top of the sensing layer which in turn ultimately hampers recovery and repeatability. This can be overcome by heating the material or exposing it to UV light. Continuous UV light exposure in long term degrades the performance of the sensor.

The ultimate objective of any type of detection method is to detect the particle in subatomic level if possible to individual quantum level, so that the target of the detection can be resolved. In the case of chemical sensors, the quantum is one atom or molecule. Solid-state sensors can be categorized as solid electrolyte sensors in which sensing is in the form of conduction of ions, catalytic effect sensors where the sensing is done by the resistance change due to the effect of temperature and semiconducting oxide gas sensors which is responsible for the change in charge carrier by the reaction of molecules. Suspended graphene is chemically stable yet it interacts with the atoms present in the environment. Permutation of electronic and mechanical properties can be used to execute the transduction of the sensing signal.

When gas molecules adsorbed on the graphene surface, the local change in the carrier concentration can be considered as equivalent to doping. This doping results the 2D electron system in graphene to be tuned to a 2D electron gas or 2DEG, which can be monitored electrically in a transistor-like configuration [124]. A 2DEG can be treated as a cloud of free electrons to move in two dimensions. Unlike other material, the high mobility and metal-like conductivity observed in graphene contribute to limit the background noise in transport phenomenon. By exploiting these characteristics of graphene in simple graphene FET devices, ultra-sensitive sensor has been developed to detect parts-per-billion levels of toxic gases with fast response time which has been realized experimentally by Schedin et al. [125]. The notable fact as reported by Lu et al. that cleaned graphene (i.e. typically after a treatment to remove residual lithography resist) showed a much less sensitivity [126]. Even better reduced graphene oxide is thus needed to enhance the sensitivity, chemical affinity and selectivity for proper functionalization of graphene. Any internal change in graphene may be electronic (e.g. local change in the carrier concentration) or magnetic (such as presence of magnetic field) or chemical (e.g. adsorbed gas molecule) or physical (e.g. mechanical deformation) which is capable of inducing external charges that can change the graphene sensitivity. Jensen et al. [6] struggled to reach single-atom sensitivity, an unconventional approach to electrical detection scheme to exploit the electromechanical behaviour of suspended graphene in comparison to carbon nanotube–based nanomechanical sensors. The large area and stiffness of suspended graphene sheet and its very specific Raman spectography support in favour of mechanical mass sensors [127].

Massera [128] has recently reported two different fabrication approaches of graphene-based sensor. The first sensor was developed by micromechanical exfoliation of highly oriented pyrolytic graphite (HOPG, ZYB grades) using two kinds of tape (3M magic tape and Kapton tape). An ohmic response has been observed after electrical characterizations of the drop-casted device and sensor response has been checked to analytes like NO_2 and H_2O. Another graphene-based sensor device was fabricated by the use of a colloidal dispersion of graphene sheets obtained from a solution-phase exfoliation of graphite in dimethylformamide [129].

The random motion of charge carriers due to the thermal energy limits the sensor output and also leads to intrinsic noise resulting from individual molecules. Shedin et al. [125] have developed micrometre-sized sensors made from graphene which are capable of detecting individual gas molecule attached to or detached from graphene's surface resulting the change in the local carrier concentration in graphene sheet which leads to step-like changes in resistance. This in turn results a very-high-sensitive chemical sensor with fast response time.

Shan et al. [130] have first demonstrated the graphene-based biosensor. With glucose oxidase (GOD) as an enzyme model, Shan et al. and their group constructed a novel polyvinylpyrrolidone-protected graphene/polyethylenimine-functionalized ionic liquid/GOD electrochemical biosensor.

In a work carried out by Dan et al. [131], a cleaning process was described to eliminate the contamination from the sensor device thus showing the intrinsic sensitivity of graphene-based sensors. The contamination layer was removed by a high-temperature cleaning process in a reducing (H_2/Ar) atmosphere. The theoretical aspect of graphene-based sensor has also been successfully presented by several authors who provide us a clear understanding on the effect of absorption of the gas or biomolecules on the graphene surface and their effect on the mobility of graphene and the charge transport between the molecules and the graphene surface. The effect of doping on the sensing mechanism has also been scrutinized carefully as a course of action [132,133].

The effect of contact between graphene oxide and electrodes has also been analyzed by several researchers. According to their opinion, electrode is likely to contribute to the overall sensing response because of the adsorbate-induced Schottky barrier variation. Lu et al. [134] have worked on hydrophilic graphene oxide sheets uniformly suspended in water which were first dispersed onto gold interdigitated electrodes. The partial reduction of the GO sheets was then achieved through low-temperature, multistep annealing of the device in argon ambient at atmospheric pressure. The electrical conductance of GO was deduced after each heating cycle to understand the level of reduction. The thermally reduced GO showed p-type semiconducting behaviour in ambient conditions and was responsive to low-concentration NO_2 and NH_3 gases diluted in synthetic air at room temperature.

Shafiei et al. [135] presented gas-sensing properties of Pt/graphene-like nanosheets towards hydrogen gas. The graphene-like nanosheets were produced via the reduction of spray-coated graphite oxide deposited on SiC substrates by hydrazine vapour. Current, voltage and dynamic responses of the sensors were studied towards different concentrations of hydrogen gas in a synthetic air mixture at 100°C. A voltage shift of 100 mV was recorded at 1 mA reverse bias current.

Uptake of CO_2 and CH_4 by graphenes was compared with that of activated charcoal [136], and adsorption was found to be dependent on surface areas of the studied samples, with EG (exfoliated graphene) showing the highest surface area (640 m^2/g) and SGO (SLG) showing the lowest (5 m^2/g), while activated charcoal had a surface area of 1250 m^2/g. The uptake values varied between 5 and 45 wt% in the case of graphene samples at 195 K and 0.1 MPa, with EG exhibiting the highest uptake.

Graphene sandwiched between two Au microstrips and between two Co layers shows the layer-by-layer assembly of vertically conducting graphene devices. The Au–graphene–Au junctions exhibit large magnetoresistance with ratios up to 400% at room temperature, which have potential applications in magnetic field sensors [137].

Graphene-based electrochemical sensors and biosensors are found superior among all the sensor family. Graphene is a potential electrode material in electroanalysis. Several researchers have developed [138,139] electrochemical sensors based on graphene and graphene nanocomposites.

The high electrocatalytic property of graphene towards H_2O_2 makes it as an excellent electrode material for oxidase biosensors. Graphene-based glucose biosensors have also been reported. Shan et al. [140] reported the first graphene-based glucose biosensor with graphene/polyethylenimine-functionalized ionic liquid nanocomposites. Zhou et al. [141] reported an electrochemical DNA sensor based on graphene by chemically reduced graphene oxide.

8.7 Summary

Graphene is readily available from graphite and particularly the matter of concern for developing countries to avoid the highly expensive sample growing technique. Graphene is exceptionally good in its electronic property and is a wonder material in electronics. In graphene, there is beautiful amalgamation of quantum mechanical property and particle physics. The property of graphene can also be modified with the application of electric and magnetic fields, by addition of extra layers, by tailoring its geometry and finally by chemical doping. Moreover, graphene can be easily probed by various scanning techniques from mesoscopic to atomic scales, as it is associated with simple 2D structure. This makes graphene one of the most versatile systems in condensed matter physics. So, many researches have been performed on graphene to find out its unusual nature differing from semiconductor zero density of states and metal gaplessness, but still there are many relevant questions to be answered. For instance, considering the limiting factors of electronic mobility remains an important question. Another issue that may be explored is the effect of bending in free suspended graphene. Being suspended, the back electrode alters the shape of the graphene membrane. Most popular synthesis techniques have been discussed for the ease of researchers. However, the low-cost synthesis technique with low hazard and high purity is still not mentioned in any literature. A real milestone has been achieved by fabrication monolayer graphene. The chemical exfoliation technique looks appropriate in many sense. Still, graphene sheets from chemical reduction route tend to re-stack during the synthesis and the processing [122]. Further works are suggested to stop this re-stacking and to recover the spreading of graphene in solvents [142,143]. The effect of doping with heteroatoms (nitrogen, boron, etc.) is yet to be investigated. In the case of graphene, some works have already been reported with nitrogen doping at high temperatures (600°C–1000°C) [144].

Besides the unusual basic properties, graphene has the prospect for a large number of applications. One of the most promising applications is chemical sensor/biosensor as the functionality of graphene can be altered by chemical or structural modification.

How the absorption mechanism of molecules on graphene surface, direction dependency of biomolecules on the graphene and interaction procedures with these molecules affect the transport properties of graphene can be explored further for proper understanding of graphene-based sensors. Further, the adsorption of transition metal can lead to a new hybridization effects to adapt the new electronic structure. This can influence the introduction of d and f electrons in the graphene lattice by indulging a significant enhancement of the electron–electron interactions.

Most of the research reported so far is concentrated with monolayer graphene probably due to its simplicity and curiosity of producing one-atom thick material in electronic industry. The knowledge on multilayer graphene is still to be acquired as it is a productive area of research.

In summary, graphene is an excellent material for electronics, and there is much more to be explored for scientific research and application of graphene-based devices. Just to wait for days to come.

References

1. M. Inagaki, Y. A. Kim and M. Endo, Graphene: preparation and structural perfection, *J. Mater. Chem.* 21, 3280–3294 (2011).
2. H. P. Boehm, R. Setton, and E. Stumpp, Nomenclature and terminology of graphiteintercalation compounds. *Carbon* 24, 241–245 (1986).
3. K. Novoselov, A. Geim, S. Morozov, D. Jiang, Y. Zhang, S. Dubonos, I. Grigorieva, and A. Firsov, Electric field effect in atomically thin carbon films, *Science* 306, 666–669 (2004).

4. L. D. Brown, M. Jaros, and D. Ninno, Momentum-mixing-induced enhancement of band nonparabolicity in GaAs-$Ga_{1-x}Al_xAs$ superlattices, *Phys. Rev. B* 36, 2935–2937 (1987).
5. J. Hass, W. A. de Heer, and E. H. Conrad, The growth and morphology of epitaxial multilayer graphene, *J. Phys. Cond. Matter.* 20, 323202 (27pp) (2008).
6. K. Jensen, K. Kim, and A. Zettl, An atomic-resolution nanomechanical mass sensor, *Nat. Nanotechnol.* 3, 533–537 (2008).
7. A. H. Castro Neto, F. Guinea, N. M. R. Peres, K. S. Novoselov, and A. K. Geim, The electronic properties of graphene, *Rev. Mod. Phys.* 81, 109–162 (2009).
8. M. I. Katsnelson, K. S. Novoselov, and A. K. Geim, Chiral tunnelling and the Klein paradox in graphene, *Nat. Phys.* 2, 620–625 (2006).
9. X. Du, I. Skachko, A. Barker, and E. Y. Andrei, Approaching ballistic transport in suspended graphene, *Nat. Nanotechnol.* 3, 491–495 (2008).
10. G. Van Lier, C. Van Alsenoy, V. Van Doren, and P. Geerlings, Ab initio study of the elastic properties of single-walled carbon nanotubes and graphene, *Chem. Phys. Lett.* 326, 181–185 (2000).
11. C. D. Reddy, S. Rajendran, and K. M. Liew, Equilibrium configuration and continuum elastic properties of finite sized graphene, *Nanotechnology* 17, 864–870 (2006).
12. K. N. Kudin, G. E. Scuseria, and B. I. Yakobson, C2F, BN, and C nanoshell elasticity from ab initio computations, *Phys. Rev.* B 64, 235406–235415 (2001).
13. I. W. Frank, D. M. Tanenbaum, A. M. Van Der Zande, and P. L. McEuen, Mechanical properties of suspended graphene sheets, *J. Vac.Sci. Technol.* B 25, 2558–2561 (2007).
14. M. Poot and H. S. J. Van Der Zant, Nanomechanical properties of few layer graphene membrane, *Appl. Phys. Lett.* 92, 063111 (2008).
15. C. Gomez-Navarro, M. Burghard, and K. Kern, Elastic properties of chemically derived single graphene sheets, *Nano Lett.* 8, 2045–2049 (2008).
16. Z. H. Ni, H. M. Wang, J. Kasim, H. M. Fan, T. Yu, Y. H. Wu, Y. P. Feng, and Z. X. Shen, Graphene thickness determination using reflection and contrast spectroscopy, *Nano Lett.* 7, 2758–2763 (2007).
17. S. Ghosh, I. Calizo, D. Teweldebrhan, E. P. Pokatilov, D. L. Nika, A. Balandin, W. Bao, F. Miao, and C. N. Lau, Extremely high thermal conductivity of graphene: Prospects for thermal management applications in nanoelectronic circuits, *Appl. Phys. Lett.* 92, 151911-3 (2008).
18. S. Bae, H. Kim, Y. Lee, X. Xu, J. S. Park, and Y. Zheng, Roll-to-roll production of 30-inch graphene films for transparent electrodes, *Nat. Nanotechnol.*, 5, 574–578 (2010).
19. T. Gokus, R. R. Nair, A. Bonetti, M. Bohmler, A. Lombardo, and K. S. Novoselov, Making graphene luminescent by oxygen plasma treatment, *ACS Nano* 3, 3963–3968 (2009).
20. K. I. Bolotin, K. J. Sikes, Z. Jiang, M. Klima, G. Fudenberg, J. Hone, P. Kim, and H. L. Stormer, Ultrahigh electron mobility in suspended graphene, *Solid State Commun.* 146, 351–355 (2008).
21. T. Durkop, S. A. Getty, E. Cobas, and M. S. Fuhrer, Extraordinary mobility in semiconducting carbon nanotubes, *Nano Lett.* 4, 35–39 (2004).
22. S. V. Morozov, K. S. Novoselov, M. I. Katsnelson, F. Schedin Elias, J. A. Jaszczak, and A. K. Geim, Giant intrinsic carrier mobilities in graphene and its bilayer, *Phys. Rev. Lett.* 100, 016602 (2008).

23. N. Stander, B. Huard, and D. Goldhaber-Gordon, Evidence for klein tunneling in graphene p-n junctions, *Phys. Rev. Lett.* 102, 026807 (2009).
24. P. Guo, H. H. Song, and X. H. Chen, *Electrochemical* performance of graphene nanosheets as anode, *Electrochem. Commun.* 11, 1320–1324 (2009).
25. X. Li, Y. Zhu, W. Cai, M. Borysiak, B. Han, D. Chen, R. D. Piner, L. Colombo, and R. S. Ruoff, Transfer of large-area graphene films for high-performance transparent conductive electrodes, *Nano Lett.* 9, 4359–4363 (2009).
26. K. B. Kim, C. M. Lee, and J. Choi, Catalyst-free direct growth of triangular nano graphene on all substrate, *J. Phys. Chem.* 115, 14488–14495 (2011).
27. A. K. Geim and K. S. Novoselov, The rise of graphene, *Nat. Mater.* 6, 183–191 (2007).
28. O. V. Yazyev and L. Helm, Magnetometry measurements of highly-oriented pyrolytic graphite, *Phys. Rev. B* 75, 125408–125411 (2007).
29. S. Bhowmick and V. B. Shenoy, Edge state magnetism of single. Layer graphene nanostructures, *J. Chem. Phys.* 128, 244717-1 (2008).
30. Y. Wang, Y. Huang, Y. Song, X. Y. Zhang, Y. F. Ma, J. J. Liang, and Y. S. Chen, Room-temperature ferromagnetism of graphene, *Nano Lett.* 9, 220–224 (2009).
31. V. L. J. Joly, K. Takahara, K. Takai, K. Sugihara, T. Enoki, M. Koshino, and H. Tanaka, Effect of electron localization on the edge-state spins in a disordered network of nanographene sheets, *Phys. Rev. B* 81, 115408 (2010).
32. H. Sato, N. Kawatsu, T. Enoki, M. Endo, R. Kobori, S. Maruyama, and K. Kaneko, Drastic effect of water-adsorption on the magnetism of nanomagnets, *Solid State Commun.* 125, 641–645 (2003).
33. T. Enoki and K. Takai, Unconventional electronic and magnetic functions of nanographene-based host-guest systems, *Dalton Trans.* 3773–3781 (2008).
34. C. H. Yu, L. Shi, Z. Yao, D. Y. Li, and A. Majumdar, Thermal conductance and thermopower of an individual single-wall carbon nanotube, *Nano Lett.* 5, 1842–1846 (2005).
35. S. Berber, Y. K. Kwon, and D. Tomanek, Unusually high thermal conductivity of carbon nanotubes, *Phys. Rev. Lett.* 84, 4613–4616 (2000).
36. Y. Zhu, S. Murali, W. Cai, X. Li, J. W. Suk, J. R. Potts, and R. S. Ruoff, Graphene and graphene oxide: Synthesis, properties, and applications, *Adv. Mater.* 22, 3906–3924 (2010).
37. I. K. Hsu, M. T. Pows, A. Bushmaker, M. Aykol, L. Shi, and S. B. Cronin, Optical absorption and thermal transport of individual suspended carbon nanotube bundles, *Nano Lett.* 9, 590–594 (2009).
38. W. Cai, A. L. Moore, Y. Zhu, X. Li, S. Chen, L. Shi, and R. S. Ruoff, Thermal Transport in suspended and supported monolayer graphene grown by chemical vapor deposition, *Nano Lett.*, 10, 1645–1651 (2010).
39. J. H. Seol, I. Jo, A. L. Moore, L. Lindsay, Z. H. Aitken, M. T. Pettes, X. Li, Z. Yao, R. Huang, D. Broido, N. Mingo, R. S. Ruoff, and L. Shi, Two dimensional phonon transport in supported grapheme, *Science* 328, 213–216 (2010).
40. C. Berger, Z. Song, T. X. Li, A. Y. Ogbazghi, and R. Feng, Ultrathin epitaxial graphite: 2D electron gas properties and a route toward graphene-based nanoelectronics, *J. Phys. Chem. B* 108, 19912-19916 (2004).
41. H. J. Park, J. Meyer, S. Roth, and V. Kalova, Growth and properties of few-layer graphene prepared by chemical vapor deposition, *Carbon* 48, 1088–1094 (2010).

42. Z. Chen, W. Ren, B. Liu, L. Gao, S. Pei, Z. S. Wu, J. Zhao, and H. M. Cheng, Bulk growth of mono- to few-layer graphene on nickel particles by chemical vapor deposition from methane, *Carbon* 48, 3543–3550 (2010).
43. A. Ferrari, J. Meyer, V. Scardaci, C. Casiraghi, M. Lazzeri, F. Mauri, S. Piscanec D. Jiang, K. Novoselov, S. Roth, and A. Geim, Raman spectrum of graphene and graphene layers, *Phys. Rev. Lett.* 97, 187401-1-187401-4 (2006).
44. J. Hamilton and J. Blakely, Graphene on metal surfaces, *Surf. Sci.* 91, 199–217 (1980).
45. P. Szroeder, N. G. Tsierkezos, M. Walczyk. W. Strupinski, A. Gorska-Pukownik, J. Strzelecki, K. Wiwatowski, P. Scharff, and U. Ritter, Insights into electrocatalytic activity of epitaxial graphene on SiC from cyclic voltammetry and ac impedance spectroscopy, *J. Solid State Electrochem.*, 18, 2555–2562 (2014).
46. L. P. Biro and P. Lambin, Grain boundaries in graphene grown by chemical vapor deposition, *New J. Phys.* 15, 035024 (38pp) (2013).
47. Y. Hernandez, V. Nicolosi, M. Lotya, F. M. Blighe, Z. Sun, and S. De, High-yield production of graphene by liquid-phase exfoliation of graphite, *Nat. Nanotechnol.* 3, 563–568 (2008).
48. S. Horiuchi, T. Gotou, M. Fujiwara, R. Sotoaka, M. Hirata, K. Kimoto, T. Asaka et al., Carbon nanofilm with a new structure and property, *Jpn. J. Appl. Phys.* 42, 1073–1076 (2003).
49. R. Colby, Q. Yu, H. Cao, S. S. Pei, E. A. Stach, and Y. P. Chen, Crosssectional transmission electron microscopy of thin graphite films grown by chemical vapor deposition, *Diam. Relat. Mater.* 19, 143–146 (2010).
50. Walt A. de Heer, C. Berger, X. Wu, M. Sprinkle, Y. Hu, M. Ruan, J. A. Stroscio, et al., Epitaxial graphene electronic structure and transport, *Solid State Commun.* 143, 92–100 (2007).
51. A. C. Ferrari and J. Robertson, Resonant Raman spectroscopy of disordered, amorphous, and diamondlike carbon. Resonant. *Phys. Rev. B* 64, 075414-26 (2001).
52. B. Lang, A LEED study of the deposition of carbon on platinum crystal surfaces, *Surf. Sci.* 53, 317–329 (1975).
53. Y. Wu, B. Wang, Y. Ma, Yi Huang, Na Li, F. Zhang, and Y. Chen, Efficient and large-scale synthesis of few-layered graphene using an arc-discharge method and conductivity studies of the resulting films, *Nano Res.* 3, 661–669 (2010).
54. M. Viculis, J. J. Mack, and R. B. Kaner, A chemical route to carbon nanoscrolls, *Science* 299, 1361 (2003).
55. K. S. Novoselov, D. Jiang, F. Schedin, T. J. Booth, V. V. Khotkevich, S. V. Morozov, and A. K. Geim, *PNAS* 102, 10451 (2005).
56. Y. M. Chang, H. Kim, J. H. Lee, and Y.-W. Song, Nonlinearity-preserved graphene/PVAc composite in optical deposition for fiber, *Appl. Phys. Lett.* 59, 56–59 (2010).
57. C.-L. Wong, Characterization of nanomechanical graphene drum structures, *J. Micromech. Microeng.* 20, 115029-40 (2010).
58. Z. Tang, J. Zhuang, and X. Wang, Exfoliation of graphene from graphite and their self-assembly at the oil-water interface, *Langmuir* 26, 9045–9049 (2010).
59. A. Shukla, R. Kumar, J. Mazher, and A. Balan, Graphene made easy: High quality, large-area samples, *State Commun.* 149, 718–721 (2009).
60. H. Hiura, T. W. Ebbesen, J. Fujita, K. Tanigaki, and T. Takada, Role of sp^3 defect structures in graphite and carbon nanotubes, *Nature* 367, 148–151 (1994).

61. T. W. Ebbesen and H. Hiura, Graphene in 3-dimensions: Towards graphite origami, *Adv. Mater.* 7, 582–586 (1995).
62. H. V. Roy, C. Kallinger, B. Marsen, and K. Sattler, Manipulation of graphitic sheets using a tunneling microscope, *J. Appl. Phys.* 83, 4695–4699 (1998).
63. H. V. Roy, C. Kallinger, and K. Sattler, Study of single and multiple foldings of graphitic sheets, *Surf. Sci.* 407, 1–6 (1998).
64. X. Lu, M. Yu, H. Huang, and R. S. Ruoff, Graphite with the goal of achieving single sheets, *Nanotechnology* 10, 269–272 (1999).
65. C. Gomez-Navarro, Atomic structure of reduced graphene oxide, *Nano Lett.* 10, 1144–1148 (2010).
66. I. Jung, Chemical and atomic structures of GO and rGO. *J. Phys. Chem.* C 113, 18480–18486 (2009).
67. R. A. Greinke, R. A. Mercuri, and E. J. Beck, Intercalation of graphite, U.S. Patent No. 4895713 (January 23, 1990).
68. G. H. Chen, D. J. Wu, W. U. Weng, and C. L. Wu, Exfoliation of graphite flake and its nanocomposites, *Carbon* 41, 619–621 (2003).
69. Z. S. Wu, W. C. Ren, L. B. Gao, J. P. Zhao, Z. P. Chen, B. L. Liu, D. M. Tang, B. Yu, C. B. Jiang, and H. M. Cheng, Synthesis of graphene sheets with high electrical conductivity and good thermal stability by hydrogen arc discharge exfoliation, *ACS Nano* 3, 411–417 (2009).
70. M. Lotya, Y. Hernandez, P. J. King, R. J. Smith, V. Nicolosi, L. S. Karlsson, F. M. Blighe, et al., Liquid phase production of graphene by exfoliation of graphite in surfactant/water solutions, *J. Am. Chem. Soc.* 131, 3611–3620 (2009).
71. Y. Si and E. T. Samulski, Synthesis of water soluble graphene, *Nano Lett.* 8, 1679–1682 (2008).
72. C. Nethravathi and M. Rajamathi, The production of smectite clay/graphene composites through delamination and costacking, *Carbon* 46, 1994–1998 (2008).
73. V. C. Tung, M. J. Allen, Y. Yang, and R. B. Kaner, Highthroughput solution processing of large-scale graphene, *Nat. Nanotechnol.* 4, 25–29 (2008).
74. V. C. Tung, L. Chen, M. J. Allen, J. K. Wassei, K. Nelson, R. B. Kaner, and Y. Yang, Low-temperature solution processing of graphene-carbon nanotube hybrid materials for high-performance transparent conductors, *Nano Lett.* 9, 1949–1955 (2009).
75. B. Hu, H. Ago, Y. Ito, K. Kawahara, M. Tsuji, E. Magome, K. Sumitani, N. Mizuta, K. Ikeda, and S. Mizuno, Epitaxial growth of large-area single-layer graphene over Cu (111)/sapphire by atmospheric pressure CVD, *Carbon* 50, 57–65 (2012).
76. J. Hwang, M. Kim, D. Campbell, H. A. Alsalman, J. Y. Kwak, S. Shivaraman, A. R. Woll, A. K. Singh, R. G. Hennig, S. Gorantla, M. H. Rummeli, and M. G. Spencer, van der Waals epitaxial growth of graphene on sapphire by chemical vapor deposition without a metal catalyst, *ACS Nano* 7, 385–395 (2013).
77. S. Grandthyll, S. Gsell, M. Weinl, M. Schreck, S. Hufner, and F. Muller, Epitaxial growth of graphene on transition metal surfaces: Chemical vapor deposition versus liquid phase deposition, J. Phys.: *Condens. Matter* 24, 314204 (2012).
78. H. Fukidome, Epitaxial growth processes of graphene on silicon substrates, Jpn. *J. Appl. Phys.* 49, 01AH03-1-4 (2010).
79. T. Ohta, A. Bostwick, T. Seyller, K. Horn, and E. Rotenberg, Controlling the electronic structure of bilayer graphene. *Science* 313, 951–954 (2006).
80. A. J. Vanbommel, J. E. Crombeen, and A. Vantooren, LEED and Auger electron observations of the SiC(0001) surface, *Surf. Sci.* 48, 463–472 (1975).

81. I. Forbeaux, J. M. Themlin, and J. M. Debever, High-temperature graphitization of the 6H–SiC (000(_1)) face. *Surf. Sci.* 442, 9–18 (1999).
82. I. Forbeaux, J. M. Themlin, and J. M. Debever, Heteroepitaxial graphite on 6H–SiC(0001): Interface formation through conduction-band electronic structure, *Phys. Rev. B* 58, 16396–16406 (1998).
83. A. Charrier, A. Coati, T. Argunova, F. Thibaudau, Y. Garreau, and R. Pinchaux, Solid-state decomposition of silicon carbide for growing ultra-thin heteroepitaxial graphite films, *J. Appl. Phys.* 92, 2479–2784 (2002).
84. A. Nagashima, H. Itoh, T. Ichinokawa, C. Oshima, and S. Otani, Change in the electronic states of graphite overlayers depending on thickness, *Phys. Rev. B* 50, 4756 (1994).
85. E. Rollings, G. H. Gweon, S. Y. Zhou, B. S. Mun, J. L. McChesney, B. S. Hussain, A. V. Fedorov et al., Sythesis and characterization of atomically-thin graphite films on a silicon carbide substrate, *J. Phys. Chem. Solids* 67, 2172–2177 (2006).
86. C. Soldano, A. Mahmood, and E. Dujardin, Production, properties and potential of graphene *Carbon* 48, 2127–2150 (2010).
87. P. W. Sutter, J.-I. Flege, and E. A. Sutter, Epitaxial graphene on ruthenium, *Nat. Mater.* 7, 406–411 (2008).
88. A. L. Vazquez de parga, F. Calleja, B. Borca, M. C. G. Passeggi Jr., J. J. Hinarejos, F. Guinea, and R. Miranda, Periodically rippled graphene: Growth and spatially resolved electronic structure, *Phys. Rev. Lett.* 100, 056807 (2008).
89. Q. K. Yu, J. Lian, S. Siriponglert, H. Li, Y. P. Chen, and S. S. Pei, Wafer-scale synthesis of graphene by chemical vapor deposition and its application in hydrogen sensing, *Appl. Phys. Lett.* 93, 113103 (2008).
90. H. H. Madden, J. Kuppers, and G. Ertl, Interaction of carbon monoxide with (110) nickel surfaces, *J. Chem. Phys.* 58, 3401–3410 (1973).
91. Y. Gamo, A. Nagashima, M. Wakabayashi, M. Terai, and C. Oshima, Atomic structure of monolayer graphite formed on Ni(111), *Surf. Sci.* 374, 61–64 (1997).
92. N. Osamu, T. Noriaki, and M. Yoshiyasu, Thermal decomposition of acetylene on Pt(111) studied by scanning tunneling microscopy, *Surf. Sci.* 514, 414–419 (2002).
93. R. T. Vang, K. Honkala, S. Dhal, E. K. Vestergaard, J. Schadt, E. Laegsgaard, B. S. Clausen, J. K. Norskov, and F. Besenbacher, Controlling the catalytic bond-breaking selectivity of Ni surfaces by step blocking, *Nat. Mater.* 4, 160–162 (2005).
94. Y. B. Zhang, Y. W. Tan, H. L. Stormer, and P. Kim, Experimental observation of the quantum Hall effect and Berry's phase in graphene, *Nature* 438, 201–204 (2005).
95. J. J. Chen, J. Meng, D. P. Yu, and Z. M. Liao, Fabrication and electrical properties of stacked graphene monolayers, *Sci. Rep.* 4, 5065 (2014).
96. A. K. Geim, Graphene: Status and prospects, *Science* 324, 1530–1534 (2009).
97. A. J. Pollard, R. R. Nair, S. N. Sabki, C. R. Staddon, L. M. A. Perdigao, C. H. Hsu, J. M. Garfitt et al., Formation of monolayer graphene by annealing sacrificial nickel thin films. *J. Phy. Chem.* C 113, 16565–16567 (2009).
98. X. S. Li, W. W. Cai, J. H. An, S. Kim, J. Nah, D. X. Yang, R. D. Piner et al., Large area synthesis of high quality and uniform graphene films on coppoer coils, *Science* 324, 1312–1314 (2009).

99. J. C. Shelton, H. R. Patil, and J. M. Blakely, Equilibrium segregation of carbon to a nickel (111) surface: A surface phase transition, *Surf. Sci.* 43, 493–520 (1974).
100. A. Reina, H. Son, L. Jiao, B. Fan, M. S. Dresselhaus, Z. F. Liu, and J. Kong, Transferring and identification of single- and few-layer graphene on arbitrary substrates, *J. Phys. Chem.* C 112, 17741–17744 (2008).
101. X. Li, W. Cai, L. Colombo, and R. S. Ruoff, X. Li, W. Cai, L. Colombo, and R. S. Ruoff, Evolution of graphene growth on Ni and Cu by carbon isotope labelling, *Nano Lett.* 9, 4268–4272 (2009).
102. M. G. Rybin, A. S. Pozharov, and E. D. Obraztsova, Control of number of graphene layers grown by chemical vapor deposition, *Phys. Status Solidi C 7,* 2785–2788 (2010).
103. L. Gomez De Arco, Y. Zhang, A. Kumar, and C. Zhou, Synthesis, transfer, and devices of single- and few-layer graphene by chemical vapor deposition, *IEEE Trans. Nanotechnol.* 8, 135–138 (2009).
104. M. Zhu, J. Wang, B. C. Holloway, R. A. Outlaw, X. Zhao, K. Hou, V. Shutthanandan, and D. M. Manos, A mechanism for carbon nanosheet formation, *Carbon* 45, 2229–2234 (2007).
105. K. S. Novoselov, Z. Jiang, Y. Zhang, S. V. Morozov, H. L. Stormer, U. Zeitler, and J. C. Maan, Room temperature quantum hall effect in graphene, *Science* 315, 1379 (2007).
106. K. S. Kim, Y. Zhao, H. Jang, S. Y. Lee, J. M. Kim, J. H. Ahn, P. Kim, J. Y. Choi, and B. H. Hong, Large-scale pattern growth of graphene films for stretchable transparent electrodes, *Nature* 457, 706–710 (2009).
107. A. N. Obraztsov, E. A. Obraztsova, A. V. Tyurnina, and A. A. Zolotukhin, Chemical vapor deposition of thin graphite films of nanometer thickness, *Carbon* 45, 2017–2021 (2007).
108. D. Kondo, S. Sato, and Y. Awano, Self-organization of novel carbon composite structure: Graphene multi-layers combined perpendicularly with aligned carbon nanotubes, *Appl. Phys. Express* 1, 074003 (2008).
109. J. Coraux, A. T. Ndiaye, C. Busse, and T. Michely, Structural coherency of graphene on Ir(111), *Nano Lett.* 8, 565–570 (2008).
110. S. Marchini, S. Gunther, and J. Wintterlin, Scanning tunneling microscopy of graphene on Ru(0001), *Phys. Rev. B* 76, 075429 (2007).
111. G. Nandamuri, S. Roumimov, and R. Solanki, Chemical vapor deposition of graphene films, *Nanotechnology* 21, 145604-07 (2010).
112. W. Regan, N. Alem, B. Aleman, B. Geng, C. Girit, L. Maserati, F. Wang, M. Crommie, and A. Zettl, A direct transfer of layer-area graphene, *Appl. Phys. Lett.* 96, 113102 (2010).
113. C. Berger, Z. Song, X. Li, X. Wu, N. Brown, C. Naud, D. Mayou, T. Li, J. Hass, A. N. Marchenkov, E. H. Conrad, P. N. First, and W. A. de Heer, Electronic confinement and coherence in patterned epitaxial graphene, *Science* 312, 1191–1196 (2006).
114. D. A. Dikin, S. Stankovich, E. J. Zimney, R. D. Piner, G. H. B. Dommett, G. Evmenenko, S. T. Nguyen, and R. S. Ruoff, Preparation and characterization of graphene oxide paper, *Nature* 448, 457–460 (2007).
115. X. Li, W. Cai, J. An, S. Kim, J. Nah, D. Yang, R. Piner, A. Velamakanni, I. Jung, E. Tutuc, S. K. Banerjee, L. Colombo, and R. S. Ruoff, Large-area synthesis of high-quality and uniform graphene films on copper foils, *Science* 324, 1312–1314 (2009).

116. R. Ruoff, Graphene: Calling all chemists, *Nat. Nanotechnol.* 3, 10–11 (2008).
117. W. W. Cai, Synthesis and solid-state NMR structural characterization of C-13-labeled graphite oxide, *Science* 321, 1815–1817 (2008).
118. A. Lerf, H. Y. He, M. Forster, and J. Klinowski, A new structural model for graphite oxide. *J. Phys. Chem. B* 102, 4477–4482 (1998).
119. J. Shen, Y. Hu, M. Shi, X. Lu, C. Qin, C. Li, and M. Ye, Fast and facile preparation of graphene oxide and reduced graphene oxide nanoplatelets, *Chem. Mater.* 21, 3514–20 (2009).
120. L. Zhang, J. Liang, Y. Huang, Y. Ma, and Y. Wang, Size-controlled synthesis of graphene oxide sheet using chemical exfoliation, *Carbon* 47, 3365–3368 (2009).
121. H. Y. He, J. Klinowski, M. Forster, and A. Lerf, A new structural model for graphite oxide, *Chem. Phys. Lett.* 287, 53–56 (1998).
122. S. Park and R. S. Ruoff, Chemical methods for the production of graphenes, *Nat. Nanotechnol.* 4, 217–224 (2009).
123. E.-A. Kim and A. H. Castro Neto, Graphene as an electronic membrane, *Europhys. Lett.* 84, 57007 (2008).
124. T. O. Wehling, K. S. Novoselov, S. V. Morozov, E. E. Vdovin, M. I. Katsnelson, and A. K. Geim, Resonant scattering by realistic impurities in graphene. *Nano Lett.* 8, 173–177 (2008).
125. F. Schedin, A. K. Geim, S. V. Morozov, E. W. Hill, P. Blake, and M. I. Katsnelson, Detection of individual gas molecules adsorbed on graphene, *Nat. Mater.* 6, 652–655 (2007).
126. G. H. Lu, L. E. Ocola, and J. H. Chen, Gas detection using low-temperature reduced graphene oxide sheets, *Appl. Phys. Lett.* 94, 083111 (2009).
127. N. L. Rangel and J. A. Seminario, Graphene terahertz generators for molecular circuits and sensors, *Phys. Chem. A* 112, 13699–13705 (2008).
128. M. L. Miglietta , E. Massera, S. Romano, T. Polichetti, I. Nasti, F. Ricciardella, G. Fattoruso, Chemically exfoliated graphene detects NO2 at the ppb level, *Procedia Eng.* 29, 1145–1148 (2011).
129. P. Dutta and P. M. Horn, Low-frequency fluctuations in solids: noise. *Rev. Mod. Phys.* 53, 497 (1981).
130. C. Shan, H. Yang, J. Song, D. Han, A. Ivaska, and L. Niu, Direct electrochemistry of glucose oxidase and biosensing for glucose based on graphene, *Anal. Chem.* 81, 2378–2382 (2009).
131. Y. Dan, Y. Lu, N. J. Kybert, Z. Luo, and A. T. C. Johnson, Intrinsic response of graphene vapor sensors, *Nano Lett.* 9, 1472–1475 (2009).
132. O. Leenaerts, B. Partoens, and F. M. Peeter, Adsorption of H_2O, NH_3, CO, NO_2, and NO on graphene: A first-principles study, *Phys. Rev. B* 77, 125416–125421 (2008).
133. Z. M. Ao, J. Yang, S. Li, and Q. Jiang, Enhancement of CO detection in Al doped graphene, *Chem. Phys. Lett.* 461, 276–279 (2008).
134. G. Lu, L. E. Ocola, and J. Chen, Reduced graphene oxide for roomtemperature gas sensors, *Nanotechnology* 20, 445502 (2009).
135. M. Shafiei, R. Arsat, J. Yu, K. Kalantar-zadeh, W. Wlodarski, S. Dubin, and R. B. Kaner, Pt/graphene nano-sheet based hydrogen gas sensor, *Sensors Conference IEEE*, New Zealand, 295–298 (2009).
136. G. Srinivas, Y. Zhu, R. Piner, N. Skipper, M. Ellerby, and R. Ruoff, Synthesis of graphene-like nanosheets and their hydrogen adsorption capacity, *Carbon* 48, 630–635 (2009).

137. J. J. Chen, J. Meng, Y. B. Zhou, H. C. Wu, Y.-Q. Bie, Z. M. Liao, and D. P. Yu, Layer-by-layer assembly of vertically conducting graphene devices, *Nat. Commun.* 4, 1921 (2013).
138. R. L. McCreery, Advanced carbon electrode materials for molecular electrochemistry, *Chem. Rev.* 108, 2646–2687 (2008).
139. J. Wang, Carbon nanotube based electrochemical biosensors: A review, *Electroanalysis* 17, 7–14 (2005).
140. C. S. Shan, H. F. Yang, J. F. Song, D. X. Han, A. Ivaska, and L. Niu, Direct electrochemistry of glucose oxidase and biosensing for glucose based on graphene, *Anal. Chem.* 81, 2378–2382 (2009).
141. M. Zhou, Y. M. Zhai, and S. J. Dong, Electrochemical sensing and biosensing platform based on chemically reduced graphene oxide, *Anal. Chem.* 81, 5603–5613 (2009).
142. H. Bai, Y. X. Xu, L. Zhao, C. Li, and G. Q. Shi, Non-covalent functionalization of graphene sheets by sulfonated polyaniline, *Chem. Commun.* 1667–1669 (2009).
143. Y. X. Xu, H. Bai, G. W. Lu, C. Li, and G. Q. Shi, Flexible graphene films via the filtration of water-soluble noncovalent functionalized graphene sheets, *J. Am. Chem. Soc.* 130, 5856–5857 (2008).
144. Y. Y. Shao, J. Liu, Y. Wang, and Y. H. Lin, Indentation of polydimethylsiloxane submerged in organic solvents, *J. Mater. Chem.* 19, 46–59 (2009).

9

Nanocrystalline ZnO-Based Microfabricated Chemical Sensor

9.1 Introduction

At present time, solid-state gas sensors based on semiconducting sensing materials represent a viable solution for an increasing number of applications, ranging from health care and safety to quality control in industrial processes. The necessity of sensors with improved performances leads to continuous research efforts aiming at the optimization of sensing materials, of device design and of the operating modes. Temperature is one of the main factors, which determine the sensitivity, selectivity and response time of sensors [1].

Low power consumption is a fundamental requirement for a sensor system with an acceptable battery lifetime. Conventional metal oxide gas sensors, which are commonly used for sensing inflammable gases (like CH_4) and other toxic gases (like CO_2), suffer from relatively high temperature (≥300°C) leading to high power consumption (e.g. pellistors require 350–850 mW and Taguchi gas sensors require 230–760 mW). However, the application of silicon microelectromechanical structure (MEMS) technology may permit the desired benefits of reduced thermal mass, miniaturization, low power, reproducibility and low unit cost.

Most of the papers reported so far deal with the design of either a platinum or polysilicon microheater, particularly applicable in the higher temperature range (400°C–700°C), for their better high-temperature stability and complementary metal–oxide–semiconductor (CMOS) process compatibility. Generally, metal oxide gas sensor used suspended-type microheaters on the SiO_2/Si_3N_4 composite thin membrane as their operating temperature was very high. However, this results in thermal stress causing the generation of microcracks which in turn leads to shorter lifetime and non-uniformity of temperature over the active area. This non-uniformity further leads to a reduced sensing area which degrades the sensitivity of the gas sensor.

This chapter gives the overall description of MEMS-based gas sensors based on zinc oxide to make a substantial advancement in gas sensor engineering.

9.2 Device Structure: Vertical and Horizontal

Before the evolution of MEMS, gas sensors (mainly ceramic based) have very high power consumption (of the order of 200 mW–1 W) due to their excessive thermal mass. Moreover, response time was also very high. But with MEMS and proper thermal isolation between sensor element and substrate, this power consumption has been scaled down to about only 30–150 mW. And response is also faster in case of MEMS-based gas sensors. Heater and electrode structures, which are needed to control the sensor parameters, can be integrated in two different ways. One is vertical and the other is horizontal. In the vertical structure, the output contact electrode is placed on top of the sensing layer, i.e. the sensing layer is sandwiched between the heater and electrode. In the case of the horizontal structure, the heater and electrode are placed side by side in the same layer, sharing the common active area. Figure 9.1a and b shows two types of device

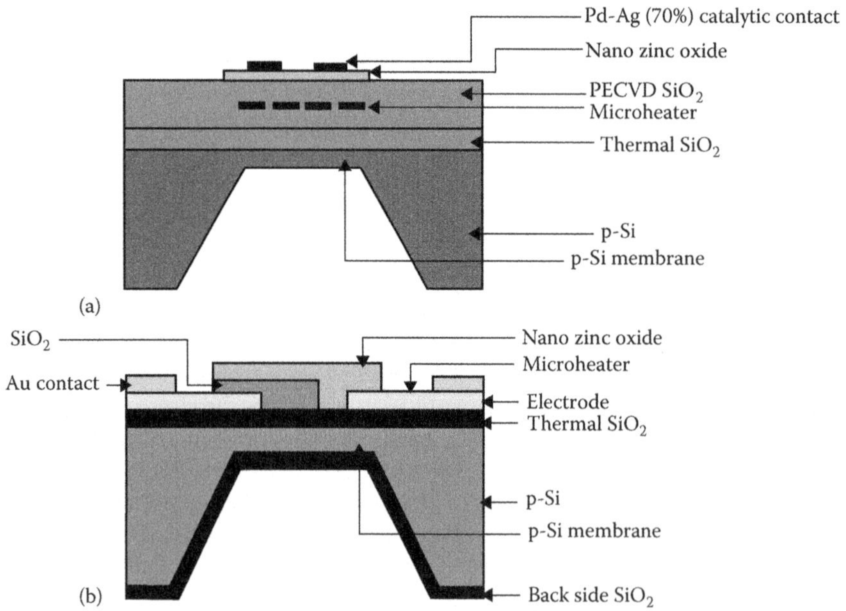

FIGURE 9.1
(a) Vertical and (b) horizontal approach.

structures. Two different types of heater design are shown in the following text for the aforementioned two approaches.

9.3 Comparison of Vertical and Horizontal Structure

The advantage of the horizontal approach is that no additional process steps are needed to form heater and electrode structures. A single lithographic step is sufficient to fabricate the device. This is a fabrication-friendly cost-effective process. As heater and electrode are sharing the same active area, the device size is comparatively larger. The robustness of the device is a prime factor as the heater and electrode both are placed on top of the thin membrane.

On the other hand, the design of structures which guarantee uniform temperature distribution as well appropriate resistance measurements is called the vertical approach. Very small active area supports much more miniaturization of the device.

9.4 Metal–Insulator–Metal Structure

Among various gas sensor structures reported so far, probably the least investigated one is the metal–(active) insulator–metal (MIM) structure [1–4]. The advantage of this device structure, first reported by Fonash et al. in 1974 [5], is its vertical electron transport mechanism which ultimately leads to improved response magnitude with fast response (compared to the conventional devices with planar configurations) and recovery for gases like CH_4 [1–4].

The most investigated sensor structure is undoubtedly the resistive planar one where the electron transport through the sensing layer is between the two electrodes placed horizontally onto the sensing layer (mostly semiconducting metal oxides) [10]. The incorporation of various catalytic noble metals (e.g. Pd, Pt), either in the form of the electrode or as a doping or modifying agent, is well established to improve the sensor performance (sensitivity, operating temperature, response time, etc.) significantly. 'In general, an oxide sensor with catalytic metal contact becomes more efficient, provided most of the electrons generated in the sensing catalytic electrode as a result of gas interaction process, can be collected by the second electrode without much carrier annihilation in the transport mechanism' [4]. The vertical structures fabricated by depositing the sensing layer on the conducting surface satisfy the aforementioned conditions to a large extent, and hence the vertical transport path has higher charge transfer efficiency; therefore high and fast response has been observed, than a relatively resistive planer sensor

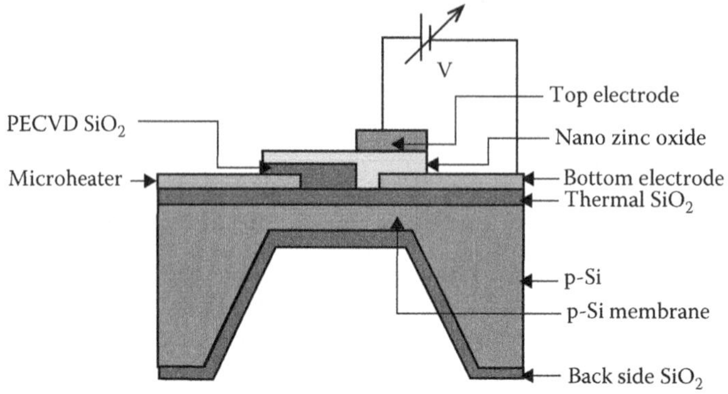

FIGURE 9.2
MEMS-based MIM structure.

structure. The potentiality of such configuration, using Pd–Ag, Pt and Rh as top catalytic electrodes and Zn as a non-catalytic bottom electrode metal and ZnO as the active insulator, was investigated thoroughly by Basu and co-workers [1–4] for gases like H_2 and CH_4. In Figure 9.2, MEMS-based MIM sensor structure has been presented.

9.5 Nanocrystalline ZnO as Sensing Material

Semiconducting metal oxides like SnO_2, ZnO and TiO_2 are by far the most popular for low working temperature though all those materials are n-type semiconductors, which have a typical working temperature in the range of 200°C–600°C [6,7]. Among them, zinc oxide (ZnO) is a well-known metal oxide semiconductor widely used in the production of many applications for its excellent properties such as piezoelectricity, transparency and chemical and thermal stability. Some of the unique features of ZnO include ease of fabrication of both n- and p-type material, high electron mobility, wide band gap, compatibility with standard CMOS technology and fairly good lattice matching with Si and SiO_2 substrate. Recently, several nanoforms of ZnO, like nanorods, nanowire, nanobelt, nanotube and nanoflakes, have been reported by various researchers [1–7] across the globe for the different fields of application including sensors owing to its inherent oxygen vacancies (ZnO1-x).

Particularly, nanocrystalline ZnO has its potential application in reducing the operating temperatures and increasing sensitivity as well as improving the response time of gas sensors due to its increased surface-to-volume ratio.

Various physical and chemical routes, such as physical vapour deposition, thermal oxidation of Zn foil, metal organic chemical vapour deposition, colloidal wetting chemical synthesis, sputtering, pulse laser ablation, sol-gel, galvanic method and chemical bath deposition (CBD), have been used to prepare a wide range of ZnO nanostructures, including novel ZnO nanoarchitectures such as nanoparticles, nanowires (NWs), nanobelts, nanotubes, nanorings, nanohelixes/nanosprings, nanobows, nanocombs and nanocages [8–12]. The physical deposition routes have the advantages of producing high-quality materials, but also the disadvantage of the need for high temperature. However, for an extensive use in commercial applications, pure ZnO films must be prepared by a low-temperature deposition methodology. In addition, the large-scale industrial preparation of multidimensional ZnO nanomaterials by a controlled methodology is also very difficult. Compared with the technologically demanding physical deposition techniques, the deposition of ZnO film from aqueous solution represents a simple and effective route. Aqueous chemical synthesis of nanostructure of ZnO is the most economical and energy-efficient method and enables good morphological control of the nanostructure [13,14]. In the wet chemical route, the basic building blocks are ions instead of atoms, and therefore the preparative parameters are easily controllable.

Depending on the fabrication process, material qualities such as impurities, their concentration, H content, number of point defects and grain boundaries can vary significantly. This chapter reviews the structural, chemical and electrical characteristic studies in the field of ZnO.

9.6 Sensing Layer Deposition by Chemical Route

There are several methods of sensing layer deposition. Few of them, depending on the ease of fabrication and cost, have been presented here.

9.6.1 Sol-Gel Method for Synthesis of ZnO Thin Films

Sol-gel is a chemical synthesis process where solvated metal alkoxides (the solution is called the 'sol') form colloidal suspensions of nanometre-sized particles (called the 'gel') via hydrolysis and subsequent polycondensation process.

For the sol-gel synthesis of ZnO thin films, zinc acetate dehydrate [$Zn(CH_3COO)_2$, $2H_2O$] and diethanolamine (DEA) [$(C_2H_4OH)_2NH$] precursors are used. The stepwise procedure is as follows:

1. Dissolve 5 g of zinc acetate powder in 50 mL isopropyl alcohol (2-propanol) in a beaker and stir with a magnetic stirrer at room temperature (~25°C) for 1 h. The solution looks milky white in colour.

2. Add slowly about 3 cc of DEA drop-wise with a syringe while stirring. The solution starts to become clearer with stirring as DEA is added. As soon as the solution turns transparent, stop adding any more drops of DEA.
3. The clear solution is again stirred at room temperature for 1 h and left for aging 24 h at room temperature.
4. After 24 h, a clear colloidal solution called gel is formed. Gel formation can be well understood by tilting the beaker to check whether the solution has become more viscous than 24 h ago.
5. This gel must be used before the solution turns opaque again due to further polycondensation which leads to network formation and thus renders the gel useless for making nanocrystalline thin films.

The reactions involved are

Hydrolysis

$$Zn(CH_3COO)_2, 2H_2O \rightarrow [Zn(CH_3COO)]^+ + (CH_3COO)^- + 2H^+ + 2(OH)^-$$

Polycondensation

$$[Zn(CH_3COO)]^+ + (C_2H_4OH)_2NH \rightarrow ZnNH(C_2H_4O)_2 + CH_3COOH + H^+$$

The $ZnNH(C_2H_4O)_2$ thus formed in the reaction eventually goes on to form ZnO.

In order to improve conductivity, aluminium doping of ZnO may be performed. For this purpose, aluminium nitrate [$Al(NO_3)_2$, $9H_2O$] has to be added to the starting materials with different concentrations. (The atomic ratio of Al/Zn in the initial solution can be varied from 0.2% to 0.9%.) The clear gel (doped or undoped) after aging is used for making thin films by the dip coating or spin coating method.

The dip-/spin-coated sample is immediately put in an oven and baked for 110°C for 10 min, thus evaporating the solvents away. The samples prepared in this way were then annealed in air at 400°C–600°C. The whole process from spin coating to annealing is repeated three times to get a thickness of about 0.9 μm of ZnO over the substrate. Such a thickness is required in order to prevent thermal degradation of the layer in subsequent repeated sensor studies at elevated temperatures ranging 100°C–300°C.

9.6.2 Chemical Bath Deposition Technique

CBD is a process in which metal cations from a metal alkoxide solvated in an aqueous medium react with OH- ions originating from another precursor (which may be strong or weak alkali like NaOH or hexamethylenetetramine [HMT]), where the pH balance of the solution may be controlled by the presence of some buffer.

9.6.2.1 Growth Mechanism

The growth mechanism of ZnO nanorods from aqueous solution involves controlled heterogeneous nucleation and homogeneous nucleation on the substrate. In the chemical solution, the hexamethylenetetramine (HMT) decomposed to formaldehyde and ammonia (NH_3), which regulate the supply of OH^-.

The overall reaction for the growth of ZnO crystals may be simplified as follows:

$$(CH_2)6N_4 + 6H_2O = 6HCHO + 4NH_3 \tag{9.1}$$

$$NH_3 + H_2O = NH_2^+ + OH^- \tag{9.2}$$

$$4OH^- + Zn^{2+} = Zn\,(OH)_4^{2-} \tag{9.3}$$

$$Zn(OH)_4^{2-} = ZnO + H_2O + 2OH^- \tag{9.4}$$

In the zinc acetate/HMT/water system, ZnO growth on the substrate is directly affected by the decomposition rate of HMT. HMT, which is extensively used in the fabrication of ZnO nanostructures, provides the hydroxide ions (OH^-) and the ammonia molecules (NH_3) to the solution as shown in Equations 9.1 and 9.2. Zn^{2+} cations form a hydroxyl complex of $Zn(OH)_4^{2-}$ anion as shown in Equation 9.3 and become the precursors of ZnO. Accordingly, the crystal nucleation and growth in the solution are controlled by interface free energy. ZnO seeds can effectively provide nucleation sites and reduce the interface energy barrier for ZnO crystal growth. Upon the hydrolysis–condensation reaction, the reactant $Zn(OH)_4^{2-}$ diffused onto the surface of the ZnO seeds and resulted in a preferential orientation along ZnO (0001) by direct homoepitaxial growth.

9.6.3 Chemical Deposition Technique

Undoped nanocrystalline *n*-ZnO thin films are deposited on the active area by a low-cost, low-temperature chemical deposition technique. The sodium zincate can be prepared by mixing zinc sulphate ($ZnSO_4$, $7H_2O$) and sodium hydroxide (NaOH) in the ratio of (1:3) in aqueous solution and stirring at room temperature.

$$Zn(SO_4) + 2NaOH = Zn(OH)_2 + Na_2SO_4 \tag{9.5}$$

$$Zn(OH)_2 + 2NaOH = Na_2ZnO_2 + 2H_2O \tag{9.6}$$

$$ZnSO_4 + 4NaOH = Na_2ZnO_2 + Na_2SO_4 + 2H_2O \tag{9.7}$$

After having a transparent homogeneous solution, the sample has to be dipped into the sodium zincate solution, and then the substrate with a thin layer of sodium zincate will be dipped into a hot water bath, and nano ZnO is formed following the reaction

$$Na_2ZnO_2 + H_2O = ZnO + NaOH \tag{9.8}$$

The resulting nanocrystalline ZnO is grown by a sodium zincate (Na_2ZnO_2) (0.125 M) aqueous solution at room temperature. As reported by Mitra et al., a 0.125 M concentration was found to be optimized because at other concentrations, the growth rate became erratic and the ZnO deposition was non-uniform [14]; then subsequent annealing follows.

References

1. P. Bhattacharyya, P.K. Basu, H. Saha and S. Basu, Fast response methane sensor based on Pd (Ag)/ZnO/Zn MIM Structure, *Sens. Lett.* 4, 371–376 (2006).
2. P. K. Basu, P. Bhattacharyya, N. Saha, H. Saha and S. Basu, Methane sensing properties of platinum catalysed nano porous zinc oxide thin films derived by electrochemical anodization, *Sens. Lett.* 6, 1–7 (2008).
3. P. K. Basu, P. Bhattachayya, N. Saha, H. Saha and S. Basu, The superior performance of the electrochemically grown ZnO thin films as methane sensor, *Sens. Actuators B* 133, 357–363 (2008).
4. P. Bhattacharyya, P. K. Basu and S. Basu, Methane detection by nano ZnO based MIM sensor device, *Sens. Transd. J.* 10, 121–130 (2011).
5. S. J. Fonash, J. A. Roger and C. H. S. Dupuy, AC equivalent circuits for MIM structures, *J. Appl. Phys.* 45(7), 2907–2910 (1974).
6. F. DiMeo Jr., I. S. Chen, P. Chen, J. Neuner, A. Roerhl and J. Welch, MEMS-based hydrogen gas sensors, *Sens. Actuators B* 117, 10–16 (2006).
7. D. L. DeVoe, Thermal issues in MEMS and microscale systems, *IEEE Trans. Compon. Packag. Technol.* 25, 117–122 (2003).
8. M. B. Rahmani, S. H. Keshmiri, M. Shafiei, K. Latham, W. Wlodarski, J. du Plessis and K. Kalantar-Zadeh, *Sens. Lett.* 7(4), 621–628 (2009).
9. M. Choudhury, S. S. Nath, D. Chakdar, G. Gope and R. K. Nath, *Adv. Sci. Lett.* 3(1), 6–9 (2010).
10. T. J. Hsueh and S. J. Chang, *Appl. Phys. Lett.* 91, 053111 (2007).
11. V. R. Shinde, T. P. Gujar and C. D. Lokhande, *Sens. Actuators B* 123, 701–706 (2007).
12. C. D. Lokhande, A. M. More and J. L. Gunjakar, *J. Alloys Comp.* 486, 570–580 (2009).
13. A. P. Chatterjee, P. Mitra and A. K. Mukhopadhyay, Chemical deposition of ZnO films for gas sensor, *J. Mater. Sci.* 34, 4225–4231, 1999.
14. P. Mitra and H. S. Maiti, A wet-chemical process to form palladium oxide sensitiser layer on thin film zinc oxide based LPG sensor, *Sens. Actuators B* 97, 49–58, 2003.

10

Nanostructures for Volatile Organic Compound Detection

10.1 Introduction

Volatile organic compound (VOC) detection is one of the key areas of research across the globe. Functional material development to be used as a sensing material is really a critical task. Alcohol sensors are widely used in industries like petrochemical, drug and food. Several nanoforms (nanorods, nanotubes, nanowires, nanoflakes, nanotubes) of semiconducting (group II–VI) metal oxides like ZnO and TiO_2 are deposited for the use of solid-state gas sensors. VOCs, which are responsible for spoilage of fruits and vegetables during storage, are to be detected in an efficient manner.

10.2 Volatile Organic Compounds

Many varieties of VOC are commercially available but among them, only four VOCs which are primarily connected to occupational hazards have been discussed here.

Methanol is one of the prime raw materials for the large-scale production of many chemical products and materials including colours, dice, drugs, perfumes and formaldehyde which are of immense use for domestic and industrial appliances. Recently, many attempts have been taken to reduce major greenhouse gas CO_2 through methanol synthesis from CO_2. By this method, even very low concentrations of CO_2 could be captured and recycled. According to the occupational health regulation, the upper concentration limits are 200 ppm average concentration for an 8 h exposure and 250 ppm maximum concentration for short-term exposure. For industries employing the CO_2 conversion process to methanol, low-concentration (<300 ppm) methanol sensors are needed as a leak detector outside the process line, while the higher-concentration (300–3040 ppm) methanol sensors are mostly used during the methanol separation from other gases (H_2, CO_2, water) in the methanol lines and also the recycle line back to the reactor. Moreover, methanol is a potential substitute for fossil fuel resources, but the toxic nature of this VOC [4] demands timely detection at different concentration levels. Therefore, the development of methanol sensor suitable to detect methanol for a wide dynamic range is extremely desirable. Moreover, the optimum temperature of such sensors should be preferably low in order to have low power consumption and long-term stability.

Ethanol is an inflammable, colourless, slightly toxic chemical compound which is mostly found in alcoholic beverages. The widespread use of ethanol covers the chemical, medical and food industry. Precise quantitative detection of ethanol vapours is call of the day for many applications including quality improvement of wines and breathe metre development to identify drunk drivers. Depending upon the percentage of alcohol in the blood (blood alcohol content [BAC]), several hazardous effects starting from mild euphoria (0.03%–0.05% by volume) to severe central nervous system depression (0.3%–0.4% by volume) or even death (>0.5% by volume) demand the precise detection of alcohol even down to ppm levels. Country-wise BAC limitation during car driving is limited to 0.20 mg/g in Sweden, 0.50 mg/g or 0.50 mg/mL in most European nations and 0.80 mg/mL in the United Kingdom, United States and Canada.

Acetone's inflammability and toxicity may sometimes cause trouble to human health, especially in case of long-term exposure, like dizziness, dryness and inflammation. Another study reported that inhalation of high levels of acetone vapour can cause irritation (throat and lung) and contraction of the chest. According to the American Conference of Governmental Industrial Hygienists (ACGIH), the threshold limit value of acetone vapour in human surroundings should be less than 750 ppm. Acetone sensors are also becoming essential for medical diagnostics and monitoring. It has been reported that acetone concentrations in the breath of a healthy person is around 5 ppm, but this increases to 300 ppm for a patient with diabetes mellitus. Apart from this, as acetone is one of the major VOCs found during long-term storage of many vegetables and fruits, its timely detection may help in predicting the freshness (or spoilage) of vegetables and fruits. Various attempts have been made since last decade to detect acetone in human body by non-invasive way. Nevertheless, in spite of extensive research efforts, fabrication of an efficient, reliable and cost-effective chemical sensor remains a technological challenge. Recently, ZnO nanostructures were reported as a promising candidate for acetone gas sensing material.

Butanone (methyl ethyl ketone) is used as a solvent in processes involving gums, resins, cellulose acetate and cellulose nitrate which are indispensible ingredients in the production of synthetic rubber, paraffin wax, lacquer, varnishes, paint remover and glues. 2-Butanone is often found in dissolved state in water or in vapour state in the air. It is also a natural product originating from trees and often found as a VOC emerging during the long-term storage of fruits and vegetables. The exhausts of cars and trucks also release 2-butanone into the air. This VOC is a by-product of several industries including meat-packing plants, sausages and other prepared meats, rice milling, edible fats and oils, malt beverages, flavouring extracts and syrups, cigarettes, grain and field beans. Due to its volatile and inflammable nature, 2-butanone can cause threat to public safety and health. Hazardous effects from exposure to 2-butanone are irritation of the nose, throat, skin and eye. Serious health hazards in animals and human beings have been reported at high butanone concentrations like 300 ppm of 2-butanone. As per Environmental Protection Agency (EPA), when inhaled, these effects may ultimately result in problems like birth defects. Kidney damage, as a hazardous outcome of butanone intoxication, has also been reported. Generally high concentrations are not expected in the usual use of 2-butanone or in the vicinity of hazardous industrial waste sites. Though, in 2005, the U.S. EPA removed butanone from the list of hazardous air pollutants, neurological, liver, kidney and respiratory effects have been reported in chronic inhalation studies of methyl ethyl ketone in animals and humans.

10.3 Different Nanostructures

10.3.1 Fabrication Processes

Different fabrication processes have already been illustrated in Chapter 7. Few reports have been presented here. Park et al. [1] have fabricated ZnO nanowire by carbothermal reduction process. An equal amount of ZnO and graphite powders (99.9%, 325 mesh) was mixed and transferred to an Al_2O_3 boat placed in a reaction tube. In the downstream of the mixed gas flow in the tube, Si substrates coated by 5–30 Å thick Au were placed. ZnO nanowires began to form on the substrate with more than 0.1% of O_2 with Ar gas. As the relative amount of O_2 gas in the flowing gas mixture increased by more than 2%, nanostructures changed from nanowires to wire arrays, rod arrays and sheet arrays.

α-Fe_2O_3 nanotubes/nanorings with various lengths and thicknesses were synthesized by a hydrothermal treatment of $FeCl_3$ solution in the presence of $NH_4H_2PO_4$ at 220°C. By adjusting the concentration of reactants, α-Fe_2O_3 nanostructures with varied sizes were produced. A detailed process has been reported in literature [2].

Copper nitrate hydroxide is a kind of intermediate of copper products and often used as pesticide; 0.79 g NH_4HCO_3 and 2 mL OP-10 were mixed and ground for 10 min; then 2.42 g $Cu(NO_3)_2{\cdot}3H_2O$ was added and ground for another 30 min. The mixture was put aside in air at room temperature for about 4 h and then washed by distilled water at least three times. The final precipitate was dried in an oven at 60°C for 6 h [3].

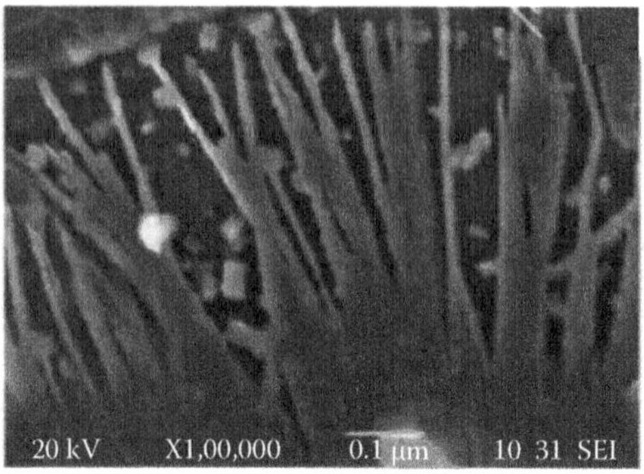

FIGURE 10.1
Side view of Ag (Silver) synthesized silicon nanowire array.

Asgar and group studied [4] the broadband optical absorption properties of silicon nanowire films (as shown in Figure 10.1) fabricated by electroless metal deposition technique followed by $HF/Fe(NO_3)_3$ solution-based chemical etching at room temperature on p-type silicon substrates. They found higher absorption than that of solid thin films of equivalent thickness. They clarified the observed behaviour of the film accordingly.

According to Hassan et al. [5], $Zn(OH)_2$–PVA nanocomposites were synthesized by the complexation of Zn ions with the PVA backbone chain, and these nanocomposites were spin coated onto a well-cleaned sapphire substrate. After annealing, ZnO nanowires are formed.

10.3.2 Characterization Processes

Different characterization processes have already been discussed in Chapter 7.

10.4 Sensing Mechanism

The basic gas sensing mechanism is the same as described in Chapter 7, i.e. a surface-related phenomenon.

Upon exposure to ethanol or any other VOCs, sensor resistance initially decreased due to the release of free electrons and then got saturated, while on cutting off the ethanol supply, the resistance increased and returned almost to its baseline value. However, sensor resistance does not exactly reach the initial (baseline) value probably due to the fact that some vapour molecules remaining adsorbed on the ZnO surface. The response, response time and recovery time are strong functions of ethanol concentrations. Increase in ethanol concentration provides more vapour molecules to be adsorbed on the sensing layer surface per unit time thereby favouring the fast electron transport kinetics; as a result, the response magnitude (RM) increases and the response time decreases. On the contrary, the recovery is slower due to slow desorption kinetics of the ethanol vapour from the interface at higher concentration regime.

The possible reason for the reduction in operating temperature due to incorporation of nanostructured sensing layer in combination with catalytic noble metal electrode and sensitizers may be attributed to the low-temperature dissociation of ethanol molecules on such sensing layers. The nanostructures, owing to its high surface-to-volume ratio, have excessive free surface energy, which ultimately leads to more adsorption per unit area of target molecules producing higher adsorption coefficient of gases or vapours. The incorporations of catalytic metal (electrode and surface modifier) further aid this ethanol dissociation mechanism.

10.5 Measurement Technique

A simple gas measurement set-up for VOC sensing is presented here (Figure 10.2). The acetone vapours were introduced in the chamber by bubbling IOLAR (purity should be 99.999%) grade N_2 (carrier gas) through the acetone kept in a conical bottle. Assuming that the carrier got saturated with acetone, the desired concentrations of the vapour were obtained in the sensing chamber by controlling the flow rate of N_2 through acetone and adding IOLAR grade N_2 (diluents) through a separate gas flow line. The homogeneous mixture carrying the desired percentage of the target

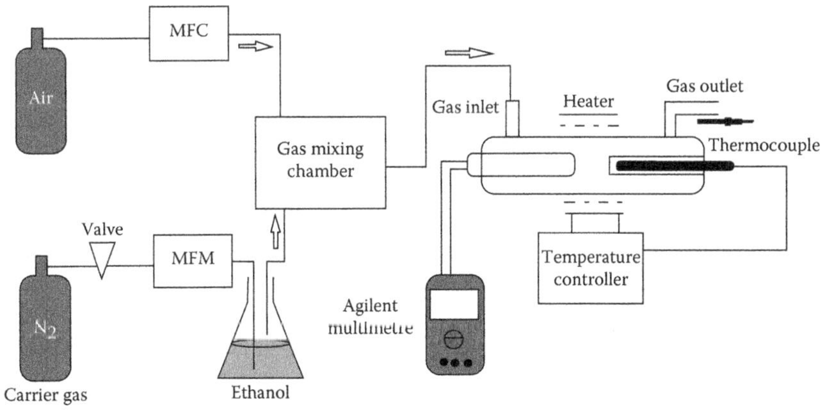

FIGURE 10.2
Gas measurement set-up for VOC sensing.

vapour was fed into the chamber with a flexible PVC pipe. For sensor resistance measurement, the electrodes are connected to an Agilent U34411A multimetre with Agilent GUI data logging software (v2.0).

The final concentration of acetone vapours was calculated using the following formula:

$$C = \frac{\dfrac{P \times L}{760 - L}}{\dfrac{P \times L}{760 - L} + L + L'} \times 10^6$$

where

L and L' are the gas flow rates of N_2 (through the bubbler) and air, respectively

P is the vapour pressure of the acetone (in mm of Hg) at room temperature (27°C)

For finding out the optimum operating temperature of sensing, sensor resistance was measured in the presence of air and acetone vapour, and the corresponding RM was calculated as a function of temperature. RM of the sensor was calculated using the following formula:

$$RM = \frac{R_a - R_g}{R_a} \times 100\%$$

where

R_g is the sensor resistance in the test gas (i.e. acetone vapour mixed with pure N_2)

R_a is the sensor resistance in air

10.6 Effect of Relative Humidity on VOC Detection

Technically, humidity is the measure of water vapour present in a gas. In general, determining the humidity in presence of gas or vapour is not an easy task. Many of the instruments have poor accuracy, narrow bandwidth, contamination problems and hysteresis. To counter the problem, there are few highly precise devices for humidity measurement needed for use in sensor calibration. The hygrometric method to measure the relative humidity (RH) in sensing gases is the most common. The instruments are generally compact, reliable and inexpensive. Hygrometric humidity sensors provide an output that indicates the humidity directly. Recent trends in metal oxide thin-film and micromachining technology make it possible to produce high-quality resistive and capacitive hygrometric sensors. The materials used to fabricate these sensing elements have the capability to change their electrical characteristics with the adsorption of water.

Atmospheric humidity affects the performance parameters like RM, response time and recovery time of semiconducting metal oxide–based vapour sensors. The water vapour in atmosphere is one of the chemisorbed species. It has been well accepted that the electrical property alteration is the outcome of charge transfer between the sensing layer and the chemisorbed species. For metal oxide semiconductor–based gas sensors, many investigations have been done, and the results show that the conductivity and sensitivity of sensors depend on the RH in the atmosphere.

However, it was observed that the presence of RH during vapour sensing significantly controls the response of such sensors.

References

1. J.-H. Park and Y.-J. Choi, Evolution of nanowires, nanocombs, and nanosheets in oxide semiconductors with variation of processing conditions, *J. Eur. Ceram. Soc.* 25, 2037–2040 (2005).
2. X. Liu, M. Zheng, Y. Lv, J. Fang, C. H. Sow, H. Fan and J. Ding, Large-scale synthesis of high-content Fe nanotubes/nanorings with high magnetization by H2 reduction process, *Mater. Res. Bull.* 48, 5003–5007 (2013).
3. X.-B. Wang and L.-N. Huang, A novel one-step method to synthesize copper nitrate hydroxide nanorings, *Trans. Nonferrous Met. Soc.* 19, 480–484 (2009).
4. M. A. Asgar, M. Hasan, M. F. Huq, and Z. H. Mahmood, Broadband optical absorption measurement of silicon nanowires for photovoltaic solar cell applications, *International Nano Letters* 4, 1–5 (2014).
5. J. J. Hassan, M. A. Mahdi, C. W. Chin, H. Abu-Hassan and Z. Hassan, A high-sensitivity room-temperature hydrogen gas sensor based on oblique and vertical ZnO nanorod arrays, *Sens. Actuators B* 176, 360–367 (2013).

11

Sensor Interfaces

11.1 Signal Processing

A sensor system has become very important to make us aware about our surroundings and to provide security and surveillance in respect of both our health and environment, whatever may be the type of sensor. Signal-conditioning unit is an essential part of a gas sensor system as otherwise it is not possible to convey the information to the common people. Especially field engineers are mostly benefited as they have to move in hazardous environments at any point in time. This unit not only detects but also determines the concentration level of a particular gas. In coal mines application, it can include an additional feature of safety and bring the reliability. Where the system is highly automated, this can serve as a monitoring device and give feedback [1–10] through the control circuitry. Integrated sensor application is always preferable for hazardous environments like underground coal mines where continuous monitoring of hazardous gas is extremely required, and it is always preferable to have the signal-conditioning unit present on the same chip as per modern sensor technology so that any type of adverse situation can be taken care of [11,12]. Several reports are there on such on-chip signal processing unit of gas sensor. For instance, as per the report of Graf et al. [13], microelectromechanical systems (MEMS)-based gas sensor and necessary signal-conditioning circuits are on the same chip. They used logarithmic convertor circuits for linearizing the output voltage with respect to gas concentration.

MEMS is the heart of microfabrication technology. Over the last few decades, many authors are working on gas sensor technology for the miniaturization of the device, lower power consumption, faster response and higher sensitivity [2]. Electronic processing techniques can also be used to combine the responses of the different gas sensors of a gas sensor array by reducing or cancelling the effect of the other interfering stimulus and considering only the causative stimulus to provide a sensor output that is related to the concentration of the target gas.

Conventional signal processing techniques can be used for this purpose, for instance, the various responses of the sensing electrodes may be digitized and subsequently processed or combined in analogue form. There are obviously some reports on the development of integrated gas sensor systems with log inverter circuits for linearizing the output voltage [13].

Few reports have been found on microcontroller-based discrete signal conditioning of MEMS-based gas sensors. Oscillator circuits have also been reported to be integrated with the gas sensor for conversion of voltage to frequency providing improved resolution [14].

11.2 Smart Sensors

A tremendous advancement in the field of sensor technology gives birth of smart sensor systems. Very precisely, a smart sensor is an integrated system with the combination of a sensing element with processing capabilities provided by a microprocessor. The sensor signal is fed to the microprocessor, which processes the data and provides the information to an external user. A smart sensor not only provides customized outputs but also improves sensor performance with embedded intelligence. Multiple sensors can be included in a single smart sensor system.

It has several functional layers which perform individually the signal detection, signal processing, data validation and interpretation and signal display. The data acquisition layer performs all the conversion operation, e.g. from analogue to digital. It also receives additional parameters for the compensation of thermal drift, drift in the long-term stability, etc. With its intelligence, it validates the data being provided. The data are nothing but the information and can be transmitted to external users.

Several intelligent features have been included at the sensor level including self-calibration; self-healing corrective measures, e.g. temperature, pressure and relative humidity correction; and also self-compensated measurements, e.g. auto zero. This can now be concluded that the smart sensor system can optimize the performance of the individual sensors and guide to a better understanding of the data.

A second major implication of smart sensors is the output from a number of sensors which can be correlated not only to verify the data from individual sensors but also to provide a better situational awareness. A communication hub between individual smart sensors obviously required for that purpose. A strong coordination is possible with this networking capability.

A non-intrusive property has also been incorporated in the smart sensor system to provide the information to the user wherever and whenever and whatever is needed.

Smart sensor is recognized as smart in terms of compatibility. The output is simply compatible with the ending device. A large group of user is benefited by this sensor. It is basically a microsensor suitably integrated with proper microelectronics. This sensor comprises of an analogue signal processing unit along with digital signal processing (optional) and other circuits.

The output of a signal-conditioning unit can be either digital or analogue or quasi-digital (pulse width or pulse frequency modulation). The levels of integration can vary from device to device. As, for example,

Minimum level: Smart sensor with analogue output (Figure 11.1)

Lower level: Smart sensor with quasi-digital output (Figures 11.2 and 11.3)

Medium level: Smart sensor with digital output (Figure 11.4)

High level: Smart intelligent sensor (Figure 11.5)

Extreme level: Smart network sensor (Figure 11.6)

Network sensor delivers the processed data from the network port. No additional interface circuitry required in this system.

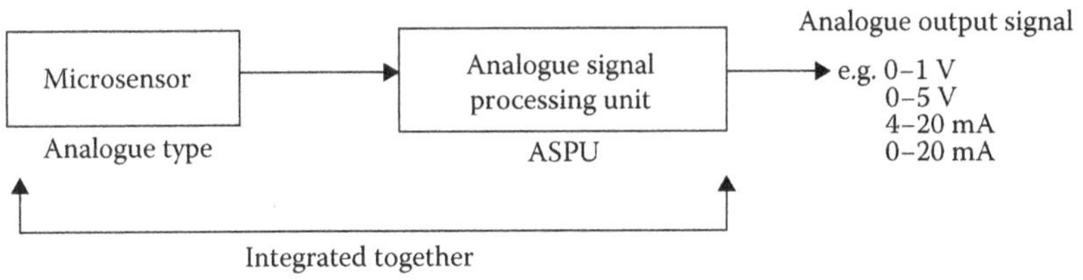

FIGURE 11.1
Smart sensor with analogue output.

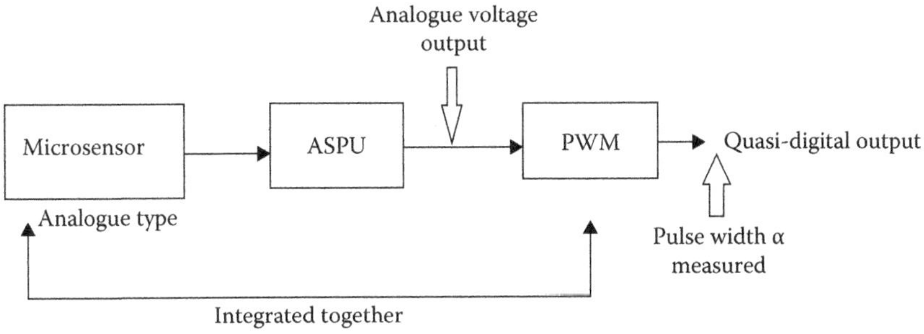

FIGURE 11.2
Smart sensor with pulse width modulation output.

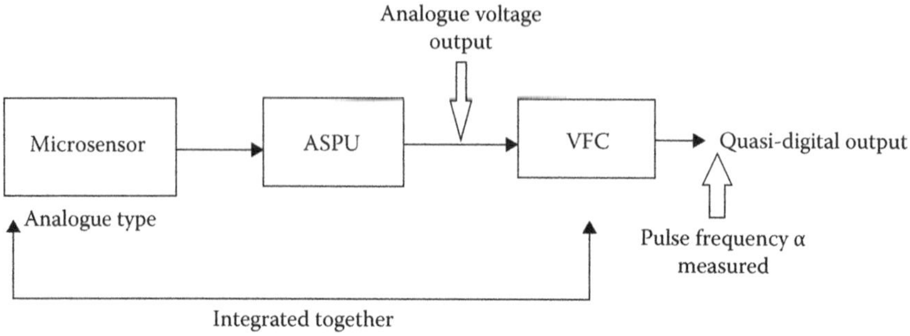

FIGURE 11.3
Smart sensor with pulse frequency output.

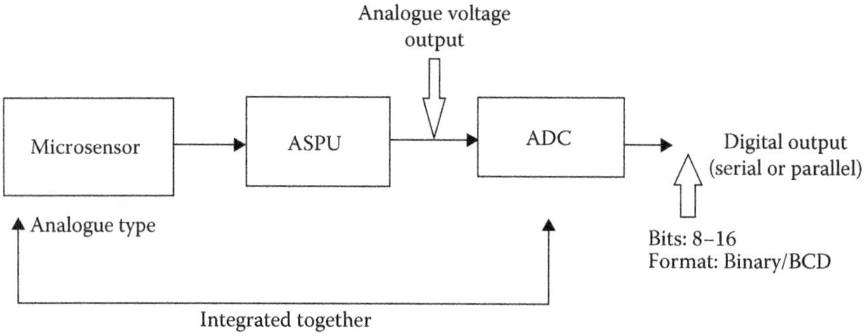

FIGURE 11.4
Smart sensor with digital output.

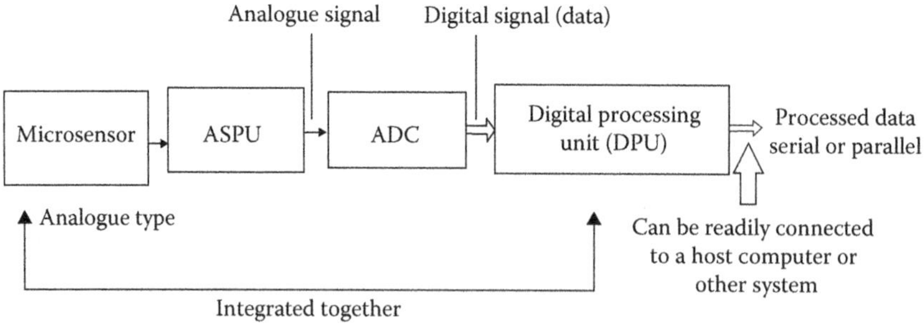

FIGURE 11.5
Intelligent sensor.

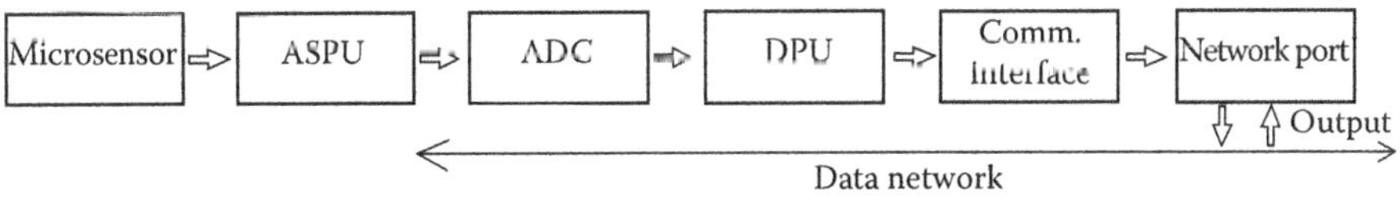

FIGURE 11.6
Network sensor.

11.2.1 System Components

The components of a smart sensor system as depicted in Figure 11.7 include sensors, power, communication and signal processing circuitry typically provided by a microprocessor. There are a lot of developments in the microprocessor technology which is beyond the scope of our discussion, but recent advancement on sensor technology enables the systems to function remotely on very little power. The ultimate objective is to have a smart sensor system with low power consumption and self monitoring capability that can sustain in the long run.

Smart sensor system is less expensive, reliable, reconfigurable and self-monitoring, and can operate for a long period of time. Microprocessor technology has been revolutionized due to the advent of microfabrication process. Microelectronics and nanotechnology play an important role in the development of smart sensor systems.

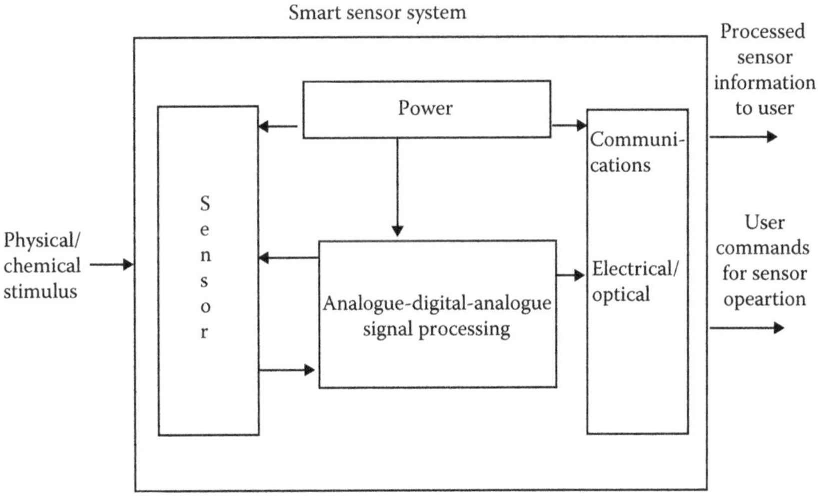

FIGURE 11.7
Block diagram of smart sensor system.

11.3 Interface Systems

The main advantage of MEMS-based gas sensor is its lower power consumption with acceptable battery lifetime particularly useful in field application. As the low power consumption is highly demanded, it is also required to use the interfacing circuit that consumes less power.

Sensing applications as designed so far require several analogue and digital blocks, such as excitation circuits, an analogue front end consisting of signal-conditioning and filtering circuitry, analogue-to-digital conversion, digital signal processing and communication blocks. Figure 11.8 illustrates the basic building blocks of a sensing interface.

Analogue and digital counterpart fabricated on the same chip not only improves the performance but also makes the system cost-effective. Such types of interfaces are known as smart sensor interfaces.

Some sensors exhibit variations in sensitivity due to the variation of different parameters which can be compensated technically by using some internal software or hardware techniques. For any type of compensation, the basic requirement is a constant current source. A constant current source basically acts as an excitation source for the sensor. The sensing interface with multichannel excitation by the constant current source is shown in Figure 11.2. Some sensing applications, such as capacitive sensing, do not require signal-conditioning circuitry, and the sensor interface should be able to directly read the sensor outputs to the analogue-to-digital converter. Above all, the sensor interface should be flexible to accommodate different sensors and signal-conditioning circuitry, if required.

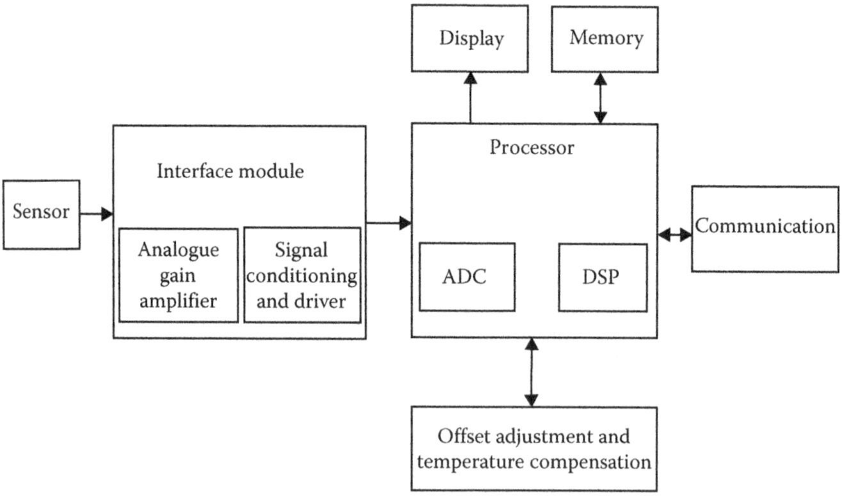

FIGURE 11.8
Basic building blocks of a sensing interface.

A smart sensor interface integrated with a microcontroller (MCU) helps the designers to develop an integrated approach. These MCUs are designed for a wide range of sensing applications. All important functions, such as analogue-to-digital conversion, communication and an excitation source, are integrated into a single chip.

References

1. C. Hagleitner, D. Lange, N. Kerness, A. Kummer, W. H. Song, A. Hierlemann, O. Brand and H. Baltes, CMOS single chip multisensor gas detection system, *Proceedings of the IEEE Conference on MEMS*, Las Vegas, NV, 2002.
2. D. W. Kensall, Integrated sensors, MEMS, and microsystems: Reflections on a fantastic voyage, *Sens. Actuators A* 136, 39–50 (2007).
3. K. D. Mitzner, J. Strnhagen and D. N. Glipeau, Development of micromachined hazardous gas sensor array, *Sens. Actuators B* 93, 92–99 (2003).
4. J. S. Suehle, R. E. Cavicchi, M. Gaitan and S. Semancik, Tin oxide gas sensor fabricated using CMOS micro-hotplates and in situ processing, *IEEE Electron Device Lett.* 14, 118–120 (1993).
5. S. Semancik, R. E. Cavicchi, M. Gaitan and J. S. Suehle, US Patent 5,345,213 (1994).
6. Figaro Products Catalogue, Figaro gas sensors 2000-series, Figaro Engineering Inc., European Office, Dusseldorf, Germany, 2006.
7. S. Fung, Z. Tang, P. Chan, J. Sin and P. Cheung, Thermal analysis and design of a micro hotplate for integrated gas-sensor applications, *Sens. Actuators A* 54, 482–487 (1996).
8. A. Gotz, I. Garca, C. Cane, E. Lora-Taniayo, M. Horrillo, G. Getino, C. Garca and J. Gutierrez, A micromachined solid state integrated gas sensor for the detection of aromatic hydrocarbons, *Sens. Actuators B* 44, 483–487 (1997).
9. G. Sberveglieri, W. Hellmich and G. Muller, Silicon hotplates for metal oxide gas sensor elements, *Microsyst. Technol.* 3, 183–190 (1997).
10. P. Bhattacharyya, P. K. Basu, B. Mondal and H. Saha, A low power MEMS gas sensor based on nanocrystalline ZnO thin films for sensing methane, *Microelectron. Reliab.* 48, 1772–1779 (2008).
11. M. Blaschke, T. Tille, P. Robertson, S. Mair, U. Weimar and H. Ulmer, MEMS gas sensor array for monitoring the perceived car-cabin air quality, *IEEE Sens. J.* 6, 1298–1308 (2006).
12. M. Afridi, A. Hefner, D. Berning, C. Ellenwood, A. Varma, B. Jacob and S. Semancik, MEMS-based embedded sensor virtual components for system-on-a-chip (SoC), *Solid State Electron.* 48, 1777–1781(2004).
13. M. Graf, D. Barrettino, M. Zimmermann, A. Hierlemann, H. Baltes, S. Hahn, Barsan and U. Weimar, CMOS monolithic metal-oxide sensor system comprising a microhotplate and associated circuitry, *IEEE Sens. J.* 4, 9–6 (2004).
14. G. C. Cardinali, L. Dori, M. Fiorini, I. Sayago, G. Faglia, C. Perego, G. Sberveglieri, V. Liberali, F. Maloberti and D. Tonietto, A smart sensor system for carbon monoxide detection, *Analog Integr. Circ. Signal Process.* 14, 275–296 (1997).

12

MEMS- and Nanotechnology-Enabled Sensor Applications

12.1 MEMS and Nanotechnology

Microelectromechanical structure (MEMS) cannot be categorized to a specific discipline, rather the interdisciplinary nature of MEMS makes it available in every aspect of engineering and manufacturing technology including integrated circuit, mechanical engineering, materials science, electrical engineering and chemical engineering. MEMS packaging issues are also broad areas of research which deal with complexity of MEMS devices to make it available in the wide range of markets. MEMS can be found in the system ranging from automotive, medical, electronic, communication to defence applications. Current MEMS devices that we use in our everyday life include accelerometers for airbag sensors, optical switches, RF switches, microvalves, inkjet printer heads, computer disk drive read/write heads, projection display chips, blood pressure sensors, gas sensors, biosensors and many other application that are all fabricated and marketed in high-profit volumes.

As everybody knows, nanotechnology is absolutely new science that comes across the market few years ago though common people are not conscious about its presence in daily life. The application of nanotechnology can be found in many everyday items. Thousands of commercial products incorporate nanomaterials. A few examples are presented here for the sake of the reader.

First, nanotechnology defends food poisoning. Many researchers across the globe have been working to keep food fresh and safe for longer periods of time. And the result is that food manufacturers are using plastic packaging made with polymer-based nanomaterials that can detect and also abolish the effect of pesticides and other contaminants found in different foods. Sometimes different nanoparticles can serve the purpose, e.g. silver nanoparticles kill bacteria from the previously stocked food if it is spread over the packing box at the time of manufacturing.

Sometimes, nanoparticles present in the packaging keep the oxygen out so that the food can be preserved for a longer period of time.

The application of MEMS and nanotechnology is overlapping everywhere. Here is a live example of that. Researchers at the Technical University of Munich recently developed carbon nanotube–based petite sensors that can be sprayed over the packaging. These lilliput sensors detect not only the concentration of VOC's but also ammonia, carbon dioxide and nitrogen oxide. The output of the sensors is linked to a wireless device that will alert store keepers about the destruction of food and to make a profitable business. Nanoparticles also minimize carbon dioxide leakage, extending the shelf life of carbonated beverages.

We use nanoparticles by our hand every day. Is it surprising? The reality is that some toothpaste carries calcium-based nanoparticles to fill the micro-cracks present in our dental enamel, and after regular use, it battles the tooth decay and makes it cavity free. Several anti-ageing lotions and creams utilize nanotechnology. They use polymeric nanocapsules which delivers active ingredients needed to keep our skin healthy. The nanoparticles of titanium dioxide and zinc oxide are used in different cosmetics.

Waterproof clothing is another example of nanotechnology research. The shape and size of the nanomaterials can be tailored to make it useful in different aspect. Nanosphere added to the fabric surface sheds the water like a lotus leaf. Sometimes, uneven textures on the fabric surface allow less surface area available for water absorption.

Even in the sports equipment, the nanotechnology has shown its expertise to make it robust and long lasting. Tennis and badminton rackets are made with carbon nanotubes, which make them lighter and stronger because the tensile strength of multiwall carbon nanotube is 63 GPa.

Footballs and tennis balls are prepared with nanoclay to make it lighter. The light weight lowers the centre of gravity, which help to swing the ball more steadily and make it bouncy. Fullerene, another form of carbon atom, helps prevent chipping and cracking of the material where it is used due to its rounded structure. Graphene oxide is also used in different objects to make them stronger and lighter.

Though nanotechnology is a relatively new science, it already has numerous applications in everyday life, ranging from consumer goods to medicine to improving the environment.

12.2 Automotive Applications: An Elaborated Study

12.2.1 Safety

One of the first commercial and widespread devices using MEMS are the automotive airbag sensors which are in the form of a single chip containing a smart sensor, or accelerometer, that measures the rapid deceleration of a vehicle (by change in voltage) on hitting an object, followed by a signal sent by an electronic control unit to trigger and explosively fill the airbag.

MEMS has replaced the initial air bag technology using mechanical 'ball and tube'–type devices, enabling the same function by integrating an accelerometer and the electronics into a single silicon chip assembled in a tiny and economic device.

The accelerometer is essentially a capacitive or piezoresistive device consisting of a suspended pendulum proof mass/plate assembly. As acceleration acts on the proof mass, micromachined capacitive or piezoresistive plates sense a change in acceleration from deflection of the plates.

The airbag sensor is fundamental to the success of MEMS and micromachining technology. With over 60 million devices sold and in operation over the last 10 years and operating in such a challenging environment as that found within a vehicle, the reliability of the technology has been proven. An example of this success is today's vehicles – the BMW 740i has over 70 MEMS devices including anti-lock braking systems, active suspension, navigation control systems, vibration monitoring, fuel sensors, noise reduction, rollover detection, seatbelt restraint and tensioning. As a result, the automotive industry has become one of the main drivers for the development of MEMS for other equally demanding environments. Some of the leading airbag accelerometer manufacturers include Analog Devices, Motorola, SensorNor and Nippondenso.

12.2.2 Vehicle Diagnostics/Monitoring

One of the more interesting applications of MEMS is in tyre pressure monitoring, or even real-time measure of tyre pressure to ensure both safety and optimized fuel performance. Another area of MEMS application is 'engine oil monitoring' where the sensors, being isolated from the medium, must survive the elevated temperature requirements.

12.2.3 Engine/Drive Train

Electronic engine control has historically been, and is expected to be, the major application area of MEMS in automotive applications. Attempts have been made to replace the manifold absolute pressure (MAP)-based air-to-fuel ratio evaluating silicon MAP sensors with mass airflow devices. Thick-film equivalents of the currently available large-sized and expensive hot-wire anemometer devices have been introduced by Bosch in 1995. An MEMS version of this device has already been developed in labs. MEMS devices are also suited for barometric pressure recordings used to provide the engine controller to compensate a rich/lean fuel-to-air mixture.

Low-cost, MEMS-based angular rate sensor similar to that developed by C.S. Draper Lab or Delco is expected to replace the existing piezoelectric and fibre-optic-based technology used for cylinder pressure measurement required for optimizing engine performance.

MEMS devices provided a lower-cost alternative solution to exhaust gas recirculation applications accommodated by ceramic capacitive pressure sensor. MEMS devices that are isolated from the media using various techniques (e.g. isolated diaphragms) could find widespread application in continuously variable transmission.

The only known application of an MEMS device in a mechanical structure is in fuel injector nozzles. Ford has micromachined silicon to create highly uniform and circular orifices for fuel injection systems.

12.2.4 Comfort, Convenience and Security

MEMS opportunities are also available in automotive comfort, convenience and security systems. The measurement of compressor pressure in the vehicle air conditioning system offers the greatest opportunity for MEMS. Currently, other technologies (e.g. Texas instrument ceramic capacitive pressure sensor) are being used. Major developments by a number of MEMS companies are actively pursuing this large opportunity.

12.3 Home Appliances

The application of nanotechnology describes how it is being integrated in small kitchen appliances. It also explains how nanotechnology is used in lighting. It describes also how paints use a nanoguard. Entirely automated kitchen is feasible by the use of nanotechnology. The benefit of MEMS in the area of energy efficiency does not need any comment.

Nowadays, MEMS sensors are being integrated in the appliances to make it smarter, one of the prominent examples is thermopile, which is a non-contact temperature sensor and most of the home appliances consist of this sensor for its better functionality. A microwave oven uses MEMS thermopile to monitor the temperature of food. The food is defrosted and cooked at the same time; it is a super time-saving instrument.

In washing machine, silver ions released during washing; a silver ion is an atom of silver with one electron missing. It sterilizes clothes without heat. MEMS pressure sensor integrated in the system helps to regulate the water pressure.

In a dish washer instrument, an MEMS pressure sensor is used to sense the soil present, depending on that it determines the right amount of water needed to clean the dishes. Here, the use of pressure sensor behaves as a significant water saver.

Refrigerators with automatic ice cube makers also use MEMS thermopile that detects when the water is frozen to drop the cube into the ice bin.

Nanomaterials guarantee the more efficient motors for refrigerators, dishwashers and dryers. Meanwhile, ultra-high brightness LEDs give more energy-efficient lighting in the home appliances.

12.4 Aerospace

12.4.1 Turbulence Control

Turbulence, being a big RE of concern in aerospace industry over the years, reduces axial flow velocity of an engine consequently reducing the thrust of the engine. Additionally, turbulence on a wing of aircraft will induce drag. With the introduction of MEMS, micron-level flow control becomes possible. Pressure sensors embedded with MEMS technology are used to sense the small changes of pressures on the walls of the tube, and temperature sensors are used to sense the variation of temperature on the walls of the tube. By this, microactuators are able to nullify most of these changes precisely and defend the turbulence control in a tube. Thrust can be increased by precisely controlling the engine inlet and exhaust with microsensors and microactuators in MEMS.

12.5 Environmental Monitoring

To improve and monitor the environmental condition, nanotechnology is being used extensively. This includes eliminating pollutants, incorporating modern equipment in the manufacturing to reduce the production of new pollutants and so on. Potential applications include the following:

- It is proved that iron nanoparticles can clean the organic solvents that are polluting groundwater. Actually, iron nanoparticles disintegrate the organic solvent wherever present. This cleaning process reduced the cost of clean water production.
- Propylene oxide is one of the most common materials while preparing plastics, paint, detergents, etc. The use of silver nanostructures used as catalyst in turn significantly reduce the polluting by-products most likely due to its catalytic effect which mainly includes reduction of reaction temperature due to the lowering of activation energy required. Generation of polluting gases can be minimized by the use of nanotechnology.
- Carbon nanotubes are being used to make windmill blades to make it robust and lighter, and therefore the amount of electricity

generated by each windmill is greater. Therefore, the electricity generated by windmills can be increased by the use of nanotechnology. Researchers have invented that silicon nanowires array can be used to fabricate a low-cost, high-efficiency solar cells which is no doubt a trademark in the field of nanotechnology.

There are three main strategies in the trade of environmental monitoring.

1. The first is the technological revolution in the engineering of individual sensors. The main engineering tool is MEMS. This includes microfluidics, the use of nanomaterials such as graphene and carbon nanotubes with the main objective of miniaturization making sensors inexpensive.
2. The second strategy is the sensitivity improvements with fast response while detecting target gases. Much lower threshold levels of detection have been achieved by the use of nanomaterials.
3. The third strategy is the development of signal conditioning unit to convey information to the common people.

There are thousand types of environmental sensors on different physical and chemical principles with different types of outputs. Typical contaminants monitored are metals, volatile organic compounds, biological contaminants and radioisotopes.

12.6 Process Engineering

Process engineering could be described as a method for shaping raw materials into usable product forms. Process engineering is followed by product engineering, i.e. it takes place after product design is complete. It takes the information from product engineer and then plan for the execution. The route of process engineering mainly involves

- Supervising product yields and process efficiencies
- Find out the ways to increase plant efficiency
- Set standards for inspection, routine test and repair of equipment during plant shutdowns
- Redesign processes that improve product quality, trim down the operating costs, improve safety and also protect the environment from pollution

MEMS finds the application in process engineering. Although based on commonly used silicon wafer processing, the manufacturing of MEMS devices requires highly specialized equipment to create mechanical structures that are a fraction of the width of a human hair. Highly flexible exposure and coating systems as well as wafer bonding equipment are essential in the processing of MEMS.

12.7 Medical Diagnostic

The milestone achieved in drug manufacturing, drug delivery, and medical diagnostics is only by the use of nanotechnology. This is still in the learning process and to know how substances react in a different manner at the particle level. Researchers are strongly searching the different aspect of nanoscale materials to make its use in medical application. The main research areas of medical diagnostics are drug synthesis, toxicity reduction and materials optimization.

Douglas English, Philip DeShong and Sang Bok Lee [1–5] are developing promising capabilities of the drug which is targeting specific cells to attack. The drug will be carried out in a durable drug carrier. This will positively reduce the side effect.

Hamid Ghandehari [6] is also working on the same area. He has developed the nanomaterials to carry the drugs and place it directly to tumours. Other organs will not be affected by it. He is also working with polymeric biomaterials for recovering cancer gene therapy and the oral administration of drugs.

The following are a few examples.

Particles can be tailored so that they are attracted to diseased cells only allowing earlier detection of disease and direct treatment of those cells with no side effect. Nanoparticles liberated in the bloodstream adhere free radicals resulting allergies and block them. And nanosilicate applied to injuries can decrease bleeding by absorbing the water and causing the blood to clot more rapidly. The structural effect of nanoshells might be used in lasers to focus the beam towards the affected cells without destructing the healthy cells around them.

The miniaturized, disposable blood pressure sensors are probably the most significant development in the field of medical diagnostic. Pressure sensor is one of the names in the successful stories of MEMS. In addition to membrane, a piezoresistive layer is applied on the membrane surface near the edges to convert the mechanical stress into an electrical voltage. Pressure generated from the deflection of the membrane. A gel is used to separate the saline solution from the sensing element.

Latest pacemaker comprises an MEMS accelerometer to monitor the patient's motion and activity and send the signals to the pacemaker to adjust its rate accordingly.

The main area of research in nanoparticle polymer-based drug delivery is the influence of architecture, drug charging and testing of toxicity on human body.

Sang Bok Lee [4] developed nanoscale materials that are hollow, tube-shaped nanostructures acting like small capsules, guard the medicine inside the shield and prevent the effect originated from the enzymatic reactions.

Lee further developed the nanocapsules with an adding layer of magnetite material inside the nanotube attributed with magnetic properties. These properties also appear to hold the tubes at their target, providing maximum therapeutic effect. Due to the presence of magnetic materials, these structures prevent the need for bio-compatibility among components and enable the nanotubes to disperse easily in aqueous solutions. The iron oxide can be the magnetic material.

There is a tremendous achievement in drug delivery and gene therapy with the advent of protein-based polymers. The University School of Pharmacy's work demonstrates that.

The main technique is that the polymer structure is related to the gene release and transfer, for applications in cancer gene therapy. Recombinant DNA technology that allows manipulation of these polymers at the molecular level has exposed new promises for controlling drug release.

References

1. J. Park, L. H. Rader, G. B. Thomas, E. J. Danoff, D. S. English and P. DeShong, Carbohydrate-functionalized catanionic surfactant vesicles: Preparation and lectin-binding studies, *Soft Matter* 4, 1916 (2008).
2. E. A. Dias, A. F. Grimes, D. S. English and P. Kambhampati, Single dot spectroscopy of two-color quantum dot/quantum shell nanostructures, *J. Phys. Chem. C* 112, 14229–14232 (2008).
3. G. B. Thomas, L. H. Rader, J. Park, L. Abezgauz, D. Danino, P. DeShong and D. S. English, Carbohydrate modified catanionic vesicles: Probing multivalent binding at the bilayer interface, *J. Am. Chem. Soc.* 131, 5471–5477 (2009).
4. S. B. Lee, Nanotoxicology: Toxicity and biological effects of nanoparticles for new evaluation standards, *Nanomedicine* 6, 759–761 (2011).
5. B. Kong, J. H. Seog, L. M. Graham and S. B. Lee, Experimental considerations on the cytotoxicity of nanoparticles, *Nanomedicine* 6, 929–931 (2011).
6. S. Sadekar, G. Thiagarajan, K. Bartlett, D. Hubbard, A. Ray, L. D. McGill and H. Ghandehari, Poly(amido amine) dendrimers as absorption enhancers for oral delivery of camptothecin, *Int. J. Pharm.* 456, 175–185 (2013).

Index

A

Acetone, 167, 194, 197–198
AFM, *see* Atomic force microscopy (AFM)
ALD, *see* Atomic layer deposition (ALD)
American Conference of Governmental Industrial Hygienists (ACGIH), 194
Anisotropic etchant/etching
 boron, 66
 characteristics, 67–69
 MESA structure, 67
 microsensors, 67
 potassium hydroxide (KOH), 66
 silicon wafers, 66–68
Atmospheric pressure CVD (APCVD), 38–39
Atomic force microscopy (AFM), 153, 158–159
Atomic layer deposition (ALD), 31–32
Automotive applications
 comfort, convenience and security systems, 212
 electronic engine control, 211–212
 safety, 210–211
 vehicle diagnostics/monitoring, 211

B

Blood alcohol content (BAC), 194
Bulk micromachining
 advantages, 64–65
 dry etching
  advantages, 72
  anisotropic etching, 72
  isotropic etching, 72
  vs. wet etching, 65
 vs. surface micromachining, 72
 wet etching
  anisotropic etchant/etching (*see* Anisotropic etchant/etching)
  compensation structures, 70–71
  convex corner undercutting, 69–70
  dangling bond, 69, 71
  vs. dry etching, 65
  etch rate, 65
  isotropic etchants/etching (*see* Isotropic etchants/etching)
  mechanism, 65
  silicon crystal structure, 70
  undercutting, 65
Butanone, 194–195

C

Carbon nanotube (CNT), 119, 147, 214
Chemical bath deposition (CBD)
 growth mechanism, 191
 HMT, 190
 sodium hydroxide, 191
 sodium zincate, 191–192
Chemical sensors
 catalytic metals, 134–135
 crystallite size determination, 139
 definition, 128
 field emission scanning electron microscopy, 139–140
 FTIR, 141
 gas *vs.* oxygen, 132–134
 metal oxide semiconductors, 129–130
 nanocrystalline materials, 130
 oxygen adsorption, 130–132
 PL, 140–141
 qualitative analysis, 141
 Raman spectroscopy, 142–143
 receptor function, 128
 sensor reliability issues, 144
 TEM, 140
 thick- and thin-film fabrication process (*see* Thick- and thin-film fabrication process)
 transducer function, 128
 UV/VIS spectroscopy, 142
 XRD, 138–139
Chemical vapour deposition (CVD), 137
 APCVD, 38–39
 atomic hydrogen, 168

bulk growth, 168
cost, 166
epitaxial growth, 169
FET, 168
flow rates, 168
MOCVD
advantage, 41
alloy semiconductor, 40
disadvantage, 42
GaAs deposition, 42
precursor molecules, 41–42
set-up, 40–41
substrate temperature, 41
nickel thin film, 167
PECVD, 39–40, 165
PMMA, 167
remote plasma, 169
Cleanroom technology
block diagram, 4
contamination, 5–6
definition, 3
design and construction, 5
monitoring, 5
standards, 3–4
Closed-membrane type microheater, 90–92
Complementary metal oxide semiconductor (CMOS), 93, 125
Contamination control, 5
CVD, *see* Chemical vapour deposition (CVD)

D

Diffraction imaging electron microscopy, 158–159
Dirac points, 151–152, 154–155
Dry etching
bulk micromachining
advantages, 72
anisotropic etching, 72
isotropic etching, 72
vs. wet etching, 65
silicon, 23
advantages, 24
dielectric etching, 26
interconnect metal etching, 26
plasma etching, 25
reactive ion etching, 25
sputter etching, 26
vapour-phase etching, 26

E

EG, *see* Epitaxial graphene (EG)
Electron beam lithography, 57–58
Electrostatic discharge (ESD), 5
Environmental monitoring, 213–214
Environmental Protection Agency (EPA), 195
Epitaxial graphene (EG), 164–165, 174
ESD, *see* Electrostatic discharge (ESD)
Ethanol, 194, 196–197

F

FEM, *see* Finite element method (FEM)
FESEM, *see* Field emission scanning electron microscopy (FESEM)
FET, *see* Field-effect transistors (FET)
Few-layer graphene (FLG), 153, 157
chemical exfoliation, 163–164
CVD
atomic hydrogen, 168
bulk growth, 168
cost, 166
epitaxial growth, 169
FET, 168
flow rates, 168
nickel thin film, 167
PECVD, 165
PMMA, 167
remote plasma, 169
EG, 164–165
micromechanical exfoliation, 161–163
Field-effect transistors (FET), 168
Field emission scanning electron microscopy (FESEM), 139–140, 157–158
Finite element method (FEM), 100–101
FLG, *see* Few-layer graphene (FLG)
Fourier transform infrared spectroscopy (FTIR), 141
Full width at half maximum (FWHM), 139

G

Glucose oxidase (GOD), 173
GO, *see* Graphene oxide (GO)
Graphene
  AFM, 158–159
  carbon allotropes, 147–148
  CNT, 147
  definition, 147
  diffraction imaging electron microscopy, 158–159
  electronic properties, 151–152
  electronic transport, 154
  FESEM, 157–158
  GO, 169–171
  Hall effect, 154–155
  hexagonal carbon lattice, 150–151
  magnetic properties, 155–156
  mechanical properties, 152–153
  optical microscopy, 157
  optical properties, 153
  potential application
    advantages, 171
    electrochemical sensors and biosensors, 174
    fabrication approaches, 172–173
    gas-sensing properties, 173
    GOD, 173
    2DEG, 172
  Raman spectrum, 160–161
  SLG/FLG
    chemical exfoliation, 163–164
    CVD (*see* Chemical vapour deposition (CVD))
    EG, 164–165
    micromechanical exfoliation, 161–163
  sp^2 hybridization, 148
  sp^3 hybridization, 148–149
  TEM, 160
  thermal properties, 156
Graphene oxide (GO), 163, 169–171, 173, 210

H

Hall effect, 154–155
HARM, *see* High-aspect-ratio micromachining (HARM)
HEPA, *see* High-efficiency particulate air filter (HEPA)
Hexamethylenetetramine (HMT), 190–191
High-aspect-ratio micromachining (HARM)
  advantages, 82
  definition, 82
  laser micromachining, 83–84
  LIGA, 83–84
High-efficiency particulate air filter (HEPA), 5
Highly ordered pyrolytic graphite (HOPG), 162–163, 172
HMT, *see* Hexamethylenetetramine (HMT)
Home appliances, 212–213
HOPG, *see* Highly ordered pyrolytic graphite (HOPG)

I

Interdigitated electrodes (IDEs), 64, 98–99, 107–108, 125
International Sensor Technology (IST), 121
Ion beam lithography, 58–59
Isotropic etchants/etching
  characteristics, 67–69
  etch rates, 66
  HNA, 66
  limitation, 66
  silicon, 67–68
  undercutting, 65

K

Kondo effect, 155

L

Liquid-phase deposition (LPD), 164

M

Manifold absolute pressure (MAP) sensors, 211
MBE, *see* Molecular beam epitaxy (MBE)
MCU, *see* Microcontroller (MCU)

Medical diagnostics, 215–216
Membrane-to-heater ratio (MHR), 93, 99
MEMS, *see* Microelectromechanical system (MEMS)
Metal–insulator–metal (MIM), 123–124, 187–188
Metallization
  electron beam evaporation, 45
  filament evaporation, 45
  induction evaporation, 45
  ohmic contact metallization, 42–44
  Schottky contact metallization, 42–44
  sputtering, 45
  wire bonding
    ball bonding, 46–47
    ball–wedge bonding, 47
Metal organic chemical vapour deposition (MOCVD)
  advantage, 41
  alloy semiconductor, 40
  disadvantage, 42
  GaAs deposition, 42
  precursor molecules, 41–42
  set-up, 40–41
  substrate temperature, 41
Metal oxide-based gas sensors
  heterojunction, 124
  homojunctions, 124
  mixed oxides, 124–125
  resistive-type, 122
  Schottky junction, 122–124
Metal oxide semiconducting (MOS)
  capacitor, 30
  thin films, 120
  transistors, 27
Methane, 132, 134, 167–169
Methanol, 193–194
MHR, *see* Membrane-to-heater ratio (MHR)
Microcontroller (MCU), 207
Microelectromechanical system (MEMS)
  advantages, 6, 14
  applications, 9–10, 14–15
  batch processing, 6, 17
  classification, 8–9
  components, 6–7
  vs. conventional ceramic sensors, 63–64
  device fabrication, 6
  history, 7–8
  metals, 12, 14
  polymers, 12–13
  polysilicon, 12–13
  SiGe, 13
  silicon (Si), 11–12
  silicon carbide (SiC), 11–13
  silicon dioxide (SiO_2), 11–12
  silicon nitride (Si_3N_4), 11–12
Microheaters
  applications, 88–89
  array structure, 108–109
  closed-membrane type, 90–92
  design issues, 93–94
  fabrication process, 106–108
  FEM, 100
  heater geometry, 95, 97–98
  heating material, 94–96
  IDEs, 98–99
  mechanical design, 101–102
  MEMS CAD tools, 100
  microhotplate, 89
  nanocrystalline metal oxide–based solid-state sensors, 87
  power consumption
    convection, 104
    heat conduction, 104–106
    heat evacuation system, 103
    radiation, 103
  screen-printed ceramic gas sensors, 88
  semiconductor gas sensors, 88
  solid-state sensor, 87–88
  suspended-membrane type, 91–93
  temperature distribution, 100–101
  thermal response time, 102
Microhotplate, 89, 93
Micromachining
  bulk micromachining (*see* Bulk micromachining)
  definition, 64
  etch-stop technique
    advantages, 76
    boron, 77–78
    electrochemical, 78–80
    photovoltaic, 80–81
    silicon-on-insulator, 81–82

HARM
advantages, 82
definition, 82
laser micromachining, 83–84
LIGA, 83–84
surface micromachining
vs. bulk micromachining, 72
etch-back, 75–76
lift-off, 73–75
materials, 74
sacrificial layer, 72
schematic diagram, 73
stiction, 73
MIM, *see* Metal–insulator–metal (MIM)
MOCVD, *see* Metal organic chemical vapour deposition (MOCVD)
Molecular beam epitaxy (MBE), 37–38
MOS, *see* Metal oxide semiconducting (MOS)

N

Nanocrystalline ZnO
CBD
growth mechanism, 191
HMT, 190
sodium hydroxide, 191
sodium zincate, 191–192
fabrication process, 189
low power consumption, 185
MIM structure, 187–188
physical and chemical routes, 189
potential application, 188
sol-gel method, 189–190
temperature, 185
vertical and horizontal structure, 186–187
Nanosensor
high surface-to-volume ratio, 126
operating temperature, 126
sensitivity, 126
thick film, 127–128
thin film, 127
Natural non-stoichiometry, 123, 130

O

Optical microscopy, 157

P

PECVD, *see* Plasma-enhanced CVD (PECVD)
Phase-shift lithography, 60–61
Photolithography
alignment, 55–56
electron beam lithography, 57–58
exposure, 55–56
ion beam lithography, 58–59
mask layout, 53–54
optical lithographic technique, 55–56
pattern formation, 55–56
phase-shift lithography, 60–61
photoresist
adhesion, 51
applications, 54–55
bubble formation, 51–52
etch resistance, 51
negative, 52
positive, 52–53
resist adhesion factors, 50
sensitivity, 50–51
postbake, 56
prebake, 54–55
stripping method, 56–57
x-ray lithography, 59–60
Photoluminescence (PL) spectroscopy, 29, 140–141
Photoresist (PR)
adhesion, 51
applications, 54–55
bubble formation, 51–52
etch resistance, 51
negative, 52
positive, 52–53
resist adhesion factors, 50
sensitivity, 50–51
Physical vapour deposition (PVD)
ALD, 31–32
electron beam evaporation, 34
MBE process, 37–38
sputtering, 36–37
thermal evaporation, 34–35
vacuum technology, 32–33
Plasma-enhanced CVD (PECVD), 39–40, 165
Plasma oxidation, 29
Polymers, 12–13, 216

Poly(methyl methacrylate) (PMMA), 83, 167
Polysilicon, 12–13, 94
PR, *see* Photoresist (PR)
Process engineering, 214–215
Propylene oxide, 213
PVD, *see* Physical vapour deposition (PVD)

R

Raman scattering, 160–161
Raman spectroscopy, 142–143
Rapid thermal oxidation (RTO), 29
Relative humidity (RH), 198–199
Response magnitude (RM), 187, 197–198

S

Schottky junction, 122–124
Semiconductor sensors
 advantages, 119
 classification, 121–122
 disadvantages, 119
 fundamentals, 121
 MEMS gas sensor, 125–126
 metal oxide–based gas sensors
  heterojunction, 124
  homojunctions, 124
  mixed oxides, 124–125
  resistive-type, 122
  Schottky junction, 122–124
 MOS, 120
 nanomaterials, 119–120
 nanosensor
  high surface-to-volume ratio, 126
  operating temperature, 126
  sensitivity, 126
  thick film, 127–128
  thin film, 127
 solid-state chemical sensors
  catalytic metals, 134–135
  crystallite size determination, 139
  definition, 128
  field emission scanning electron microscopy, 139–140
  FTIR, 141
  gas *vs.* oxygen, 132–134
  metal oxide semiconductors, 129–130
  nanocrystalline materials, 130
  oxygen adsorption, 130–132
  PL, 140–141
  qualitative analysis, 141
  Raman spectroscopy, 142–143
  receptor function, 128
  sensor reliability issues, 144
  TEM, 140
  thick- and thin-film fabrication process (*see* Thick- and thin-film fabrication process)
  transducer function, 128
  UV/VIS spectroscopy, 142
  XRD, 138–139
Sensor interfaces
 building blocks, 206
 coal mines application, 201
 conventional signal processing techniques, 202
 electronic processing techniques, 201
 integrated sensor application, 201
 MCUs, 207
 signal conditioning unit, 201
 smart sensor
  analogue output, 203
  definition, 202
  digital output, 203–204
  features, 202
  functional layers, 202
  intelligent sensor, 203–204
  network sensor, 203, 205
  pulse frequency output, 203–204
  pulse width modulation output, 203
  system components, 205
SiC, *see* Silicon carbide (SiC)
Silicon
 dielectrics
  high-pressure oxidation, 29
  oxide properties, 29–30
  plasma oxidation, 29
  rapid thermal oxidation, 29
  silicon dioxide, 27–28
  silicon nitride, 28
 dry etching, 23
  advantages, 24
  dielectric etching, 26
  interconnect metal etching, 26
  plasma etching, 25

reactive ion etching, 25
sputter etching, 26
vapour-phase etching, 26
semiconductor, 18–19
surface contamination
airborne molecular contamination, 20
ionic contamination, 21–22
organic contamination, 20–21
particle contamination, 20
wafer cleaning, 22
wet etching, 22
disadvantages, 24
metal etching, 24
silicon dioxide etching, 23–24
wet silicon etching, 23
Silicon carbide (SiC), 11–13, 19, 164–165
Silicon dioxide (SiO_2), 11–12, 23–24, 27–28, 90
Silicon–germanium alloy (SiGe), 13, 40
Silicon nitride (Si_3N_4), 11–12, 23–24, 28, 74, 81, 90
Single-layer graphene (SLG), 152
chemical exfoliation, 163–164
CVD (*see* Chemical vapour deposition (CVD))
EG, 164–165
micromechanical exfoliation, 161–163
Smart sensor
analogue output, 203
definition, 202
digital output, 203–204
features, 202
functional layers, 202
intelligent sensor, 203–204
network sensor, 203, 205
pulse frequency output, 203–204
pulse width modulation output, 203
system components, 205
Sol-gel method, 189–190
Surface micromachining
vs. bulk micromachining, 72
etch-back, 75–76
lift-off, 73–75
materials, 74
sacrificial layer, 72
schematic diagram, 73
stiction, 73
Suspended-membrane type microheaters, 91–93

T

TEM, *see* Transmission electron microscopy (TEM)
Thermocompression bonding, 46
Thermosonic bonding, 46
Thick- and thin-film fabrication process
bulk growth
hydrothermal synthesis, 136
melt growth, 135–136
vapour phase, 136
chemical route, 138
CVD, 137
sputtering, 137–138
substrates growth, 137
Transmission electron microscopy (TEM), 140, 160, 162
Turbulence control, 213
2D electron gas (2DEG), 172

U

Ultrasonic wedge bonding, 47
Undercut ratio, 70
UV/VIS spectroscopy, 142

V

Volatile organic compound (VOC) detection
acetone, 194
butanone, 194–195
ethanol, 194
fabrication processes, 195–196
gas sensing mechanism, 196–197
measurement set-up, 197–198
methanol, 193–194
relative humidity, 198–199

W

Wet etching
bulk micromachining
anisotropic etchant/etching (*see* Anisotropic etchant/etching)

compensation structures, 70–71
convex corner undercutting, 69–70
dangling bond, 69, 71
vs. dry etching, 65
etch rate, 65
isotropic etchants/etching (*see* Isotropic etchants/etching)
mechanism, 65
silicon crystal structure, 70
undercutting, 65

silicon, 22
    disadvantages, 24
    metal etching, 24
    silicon dioxide etching, 23–24
    wet silicon etching, 23

X

X-ray diffraction (XRD), 138–139
X-ray lithography, 59–60